TRANSPORTATION ISSUES, POLICIES AND R&D

# LOCAL AND RURAL ROAD SAFETY

## EFFECTIVE PLANNING AND PRACTICES

# TRANSPORTATION ISSUES, POLICIES AND R&D

Additional books in this series can be found on Nova's website under the Series tab.

Additional e-books in this series can be found on Nova's website under the e-book tab.

TRANSPORTATION ISSUES, POLICIES AND R&D

# LOCAL AND RURAL ROAD SAFETY

## EFFECTIVE PLANNING AND PRACTICES

MITCHELL BENSON
EDITOR

nova publishers
*New York*

For permission to use material from this book please contact us:
Telephone 631-231-7269; Fax 631-231-8175
Web Site: http://www.novapublishers.com



**Library of Congress Cataloging-in-Publication Data**

ISBN: 978-1-63321-170-4

*Published by Nova Science Publishers, Inc. † New York*

# CONTENTS

# PREFACE

This book guides the development of an LRSP. It provides information to local practitioners about identifying stakeholders and partnerships needed to build support, tools to analyze data, and resources to identify safety issues and select safety strategies.

Chapter 1 – Local road practitioners across the country play a critical role in addressing crash risks at the local level and may be able to identify the specific or unique conditions that contribute to crashes within their jurisdictions. The Local Road Safety Plan (LRSP) offers a foundation for consensus and focus. It defines key emphasis areas and strategies that impact local rural roads and provides a framework to accomplish safety enhancements at the local level. The LRSP helps communities take a proactive stance in reducing and preventing local road fatalities and injuries. This document guides the development of an LRSP.

Chapter 2 – Locally-owned road safety remains a challenge to many States. Several States have shown measured success in addressing local road safety. Seven States were identified to participate in the Local Road Safety Domestic Scan, allowing a team of transportation professionals from the Federal, State, and local levels to visit and document their practices. The Domestic Scan report identifies and documents practices in the planning, programming, and implementation of efforts to improve local road safety. Practices are presented in data collection and analysis; local project identification; local project administration; funding; training and technical assistance; outreach and partnerships between State Departments of Transportation (DOTs) and local agencies. The report provides States with valuable information to launch a local road safety program or implement documented practices to improve an established program.

In: Local and Rural Road Safety
Editor: Mitchell Benson

ISBN: 978-1-63321-170-4


*Chapter 1*

# DEVELOPING SAFETY PLANS: A MANUAL FOR LOCAL RURAL ROAD OWNERS*

***Andrew H. Ceifetz, P.E., Jeffrey Bagdade, P.E., Dan Nabors, P.E., Michael Sawyer, P.E. and Kim Eccles, P.E.***

## ABSTRACT

Local road practitioners across the country play a critical role in addressing crash risks at the local level and may be able to identify the specific or unique conditions that contribute to crashes within their jurisdictions. The Local Road Safety Plan (LRSP) offers a foundation for consensus and focus. It defines key emphasis areas and strategies that impact local rural roads and provides a framework to accomplish safety enhancements at the local level. The LRSP helps communities take a proactive stance in reducing and preventing local road fatalities and injuries. This document guides the development of an LRSP.

---

* This is an edited, reformatted and augmented version of a document (Report No. FHWA-SA-12-017) prepared under contract for the Federal Highway Administration, Office of Safety, dated March 2012.

## ACRONYMS

| | |
|---|---|
| AAA | American Automobile Association |
| AASHTO | American Association of State Highway and Transportation Officials |
| ATSSA | American Traffic Safety Services Association |
| ATV | All-Terrain Vehicles |
| BIA | Bureau of Indian Affairs |
| BOTS | Bureau of Transportation Safety |
| CATS | Campus Area Transportation Study |
| CERS | Center for Excellence in Rural Safety |
| CUUATS | Champaign-Urbana Urbanized Area Transportation Study |
| DOT | Department of Transportation |
| EMS | Emergency Medical Services |
| ESSI | State Farm Embedded Safety Specialist Pilot Initiative |
| FARS | Fatality Analysis Reporting System |
| FHWA | Federal Highway Administration |
| Four (4) E's | Engineering, Enforcement, Education, and Emergency Services |
| FTA | Federal Transit Administration |
| HSIP | Highway Safety Improvement Program |
| IDOT | Illinois Department of Transportation |
| LRSAG | Local Road Safety Advisory Group |
| LRSP | Local Road Safety Plan |
| LTAP | Local Technical Assistance Program |
| MPO | Metropolitan Planning Organization |
| MUTCD | Manual on Uniform Traffic Control Devices |
| NACE | National Association of County Engineers |
| NCHRP | National Cooperative Highway Research Program |
| NHI | National Highway Institute |
| NHTSA | National Highway Traffic Safety Administration |
| RPO | Regional Planning Organization |
| SAFER | Safety Evaluation for Roadways Manual |
| SAFETEA-LU | Safe Accountable Flexible Efficient Transportation Equity Act: A Legacy for Users |
| SCIL | Selected Crash Intersection Locations |
| SHSP | Strategic Highway Safety Plan |
| TIC | Transportation Information Center |
| TTAP | Tribal Technical Assistance Program |

| | |
|---|---|
| usRAP | United States Road Assessment Program |
| WACERS | Wyoming Association of County Engineers and Road Supervisors |
| WAM | Wyoming Association of Municipalities |
| WisDOT | Wisconsin Department of Transportation |
| WRRSP | Wyoming Rural Road Safety Program |
| WYDOT | Wyoming Department of Transportation |

# EXECUTIVE SUMMARY

Approximately 60 percent of all road miles in the U.S. are non-Interstate, rural roads owned and operated by local entities, such as towns, counties, and tribal governments. In 2009, 56 percent of highway deaths occurred on rural roads and the fatality rate was 2.6 times higher in rural areas than in urban areas. These data underscore the need to systematically improve road safety in rural areas.

All States have a comprehensive safety plan that provides a framework for reducing highway fatalities and serious injuries on public roads. This Strategic Highway Safety Plan (SHSP) is a data-driven plan that establishes statewide goals, objectives, and key emphasis areas that integrate the 4 E's of Safety – engineering, education, enforcement and emergency services. The SHSP can assist local practitioners in addressing safety on local rural roads but a locally-focused plan is often needed to address the unique conditions that contribute to safety problems and to assist local practitioners in making informed safety investment decisions. These challenges faced by local agencies can be addressed through the creation of a Local Road Safety Plan (LRSP).

Local practitioners play a critical role in addressing crash risks at the local level; an LRSP provides the framework for local practitioners to take a proactive stance to identify the specific or unique conditions that contribute to crashes within their jurisdictions.

Similar to a State's SHSP, an LRSP utilizes the 4 E's to address safety issues. An LRSP provides an excellent opportunity for safety stakeholders and involved agencies at all levels of government (local, State, and Federal) to work together to align and leverage resources to address the safety challenges unique to rural roads.

Additional benefits may be realized in the process of developing an LRSP to include:

- Promoting road safety awareness.
- Developing lasting partnerships that may benefit future projects.
- Instilling or enhancing a sense of collaboration among different disciplines.
- Assisting local agencies to better leverage funding.

LRSPs have been successful in comprehensively addressing safety for a variety of local agencies throughout the U.S. The safety plans developed by these agencies have strengthened multi-disciplinary commitments to road safety. For example, Olmsted County, Minnesota prepared an LRSP that resulted in the implementation of county-wide strategies such as enhancing delineation of horizontal curves, upgrading sign and pavement markings, and providing dynamic warning signs at rural stop-controlled intersections. These and other projects initiated from identified safety strategies in the LRSP received $962,000 of Federal funds for implementation.

Addressing safety on rural roads can be challenging. The development of an LRSP can serve as a cornerstone to building a comprehensive safety program to address the safety challenges on the roadways. Depending on safety needs of the jurisdiction, the LRSP will vary in size and level of detail. This guide provides the tools necessary to start a LRSP. It provides information to local practitioners about identifying stakeholders and partnerships needed to build support, tools to analyze data, and resources to identify safety issues and select safety strategies. Worksheets and other sample materials have been provided to aid in the LRSP process. Examples of programs and experiences of other agencies have also been included to provide examples of successes in implementing LRSPs and improving safety for all road users.

# INTRODUCTION

## Purpose of This Guide

The purpose of this document is to guide local rural road practitioners in developing a Local Road Safety Plan (LRSP). The expected result of a successful LRSP is to save lives and prevent injuries resulting from motor vehicle crashes. Practitioners may be road supervisors, engineers, planners, local officials, law enforcement officers, or others who deal with transportation issues. This guide may be referenced throughout the process of developing an LRSP to assist local rural road practitioners in improving safety in a comprehensive manner, incorporating engineering, enforcement, education, and emergency services measures—the 4 E's of safety—into a framework.

This guidebook outlines the process for developing an LRSP. An LRSP is a locally- coordinated safety plan that provides a comprehensive framework for reducing highway fatalities and serious injuries on local rural roads. An LRSP is flexible and utilizes the 4 E's as appropriate to establish and gain support for an agency's local safety goals, objectives, and key emphasis areas. LRSP development is usually initiated by a transportation professional or elected official.

The purpose of an LRSP is to identify key safety needs and guide investment decisions to achieve reductions in fatalities and serious injuries on local rural public roadways. An LRSP provides an opportunity for agencies at all levels of government (local, State, and Federal) and other stakeholders to work together to align and leverage resources to address the safety challenges unique to rural roads.

## Background

Approximately 60 percent of all road miles in the U.S. are non-Interstate, rural roads owned and operated by local entities, such as towns, counties, and tribal governments.[1]

Local rural roads include facilities with two or more lanes and range from paved to unpaved roadways. The majority of these roads are two- lane, paved roads. Rural roads reflect a disproportionate number of fatalities and serious injuries. In 2009, 56 percent of highway deaths occurred on rural roads, though only 23 percent of Americans live in rural areas.[2]

Furthermore, the fatality rate (deaths per 100 million vehicle miles traveled) is 2.6 times higher in rural areas than in urban areas, underscoring the need to systematically improve road safety in rural areas.

# 1. The Importance of Developing a Local Road Safety Plan

## Local Roads in Rural Areas

Over three-fourths of all road miles in the U.S. are in rural areas. Of those three million miles of rural roads, almost 80 percent are owned and operated by local entities. In 2009, rural roads accounted for approximately 33 percent of the vehicle miles traveled in the U.S.[3], but 56 percent of fatalities.

Local roads in rural areas may have design elements that increase the risk of fatalities or serious injuries, such as inappropriately high speed limits, narrow lane widths and shoulders, steep ditches, or trees close to the roadway. Additionally, the low population density and sparse land use of rural communities can increase detection, response, and travel times for emergency services, reducing key factors in crash survivability. It typically takes more than twice as long for emergency services to arrive at a crash scene in a rural community compared to an urban community.[4]

All States already have a comprehensive safety plan. A State's Strategic Highway Safety Plan (SHSP) is a statewide-coordinated safety plan that provides a comprehensive framework for reducing highway fatalities and serious injuries on all public roads. It is a major component and requirement of the Highway Safety Improvement Program (HSIP)[5] and is developed by the State Department of Transportation (DOT) in a cooperative process with local, State, Federal, and private sector safety stakeholders. The SHSP is a data-driven, comprehensive plan that establishes statewide goals, objectives, and key emphasis areas that integrates the 4 E's – engineering, enforcement, education, and emergency services.

The purpose of an SHSP is to identify the State's key safety needs and guide investment decisions to achieve significant reductions in highway fatalities and serious injuries on all public roads. Very often a State's SHSP will include local and or rural roads as a specific emphasis area for safety improvements.

An LRSP can also be valuable for improving roadway safety. Local road practitioners across the country play a critical role in addressing crash risks at the local level and may be able to identify the specific or unique conditions that contribute to crashes within their jurisdictions. The LRSP offers a foundation for consensus and focus. It defines key emphasis areas and strategies that impact local rural roads and provides a framework to accomplish safety enhancements at the local level, whereas the SHSP prioritizes safety needs and investments at the State level. However, the State SHSP may provide inputs to the LRSP. Likewise, the LRSP can feed the SHSP process to identify local road specific safety issues. Altogether, the LRSP is a coordinated effort that assists local agencies in taking a proactive stance in reducing and preventing local road fatalities and injuries. This document guides the development of an LRSP.

## Benefits of a Local Road Safety Plan

An LRSP can be an effective tool for saving lives and reducing injuries on local rural roadways. While this is ultimately the purpose of an LRSP, there are other benefits that may be realized in the process of developing an LRSP. These benefits are summarized in Table 1.1.

**Table 1.1. Summary of the Benefits of a Local Road Safety Plan**

| Benefit | Details |
|---|---|
| **Proactive Approach** | An LRSP offers a proactive approach for local road agencies to address safety issues. An LRSP can show the public and policy makers that something is being done to systematically reduce severe crashes, thereby building trust with local government officials, key stakeholders, and the general public. |
| **Develop Partnerships** | An LRSP provides local agencies an opportunity to improve relationships with the public, stakeholders, and across governmental agencies by working through a collaborative process. Improving road safety is a benefit for everyone involved. |
| **Multi-disciplinary Cooperation** | An LRSP is a multi-disciplinary approach to addressing safety. Agencies are better able to develop more effective solutions and leverage resources by considering and coordinating engineering, enforcement, education, and emergency service strategies. |
| **Safer Roadways** | An LRSP facilitates a comprehensive approach to addressing road safety that—if successfully implemented—can lead to projects that reduce severe crashes. |
| **Safety Funding** | An LRSP with a prioritized list of improvements can help agencies better justify funding requests by documenting specific needs, particularly if they are consistent with emphasis areas and strategies identified in the State's SHSP. An LRSP also shows that an agency has done its due diligence and can help an agency compete more effectively for limited funds. |
| **Managing Liability** | An LRSP is one of several proactive risk management techniques that demonstrate an agency's responsiveness to the safety needs of the public. |

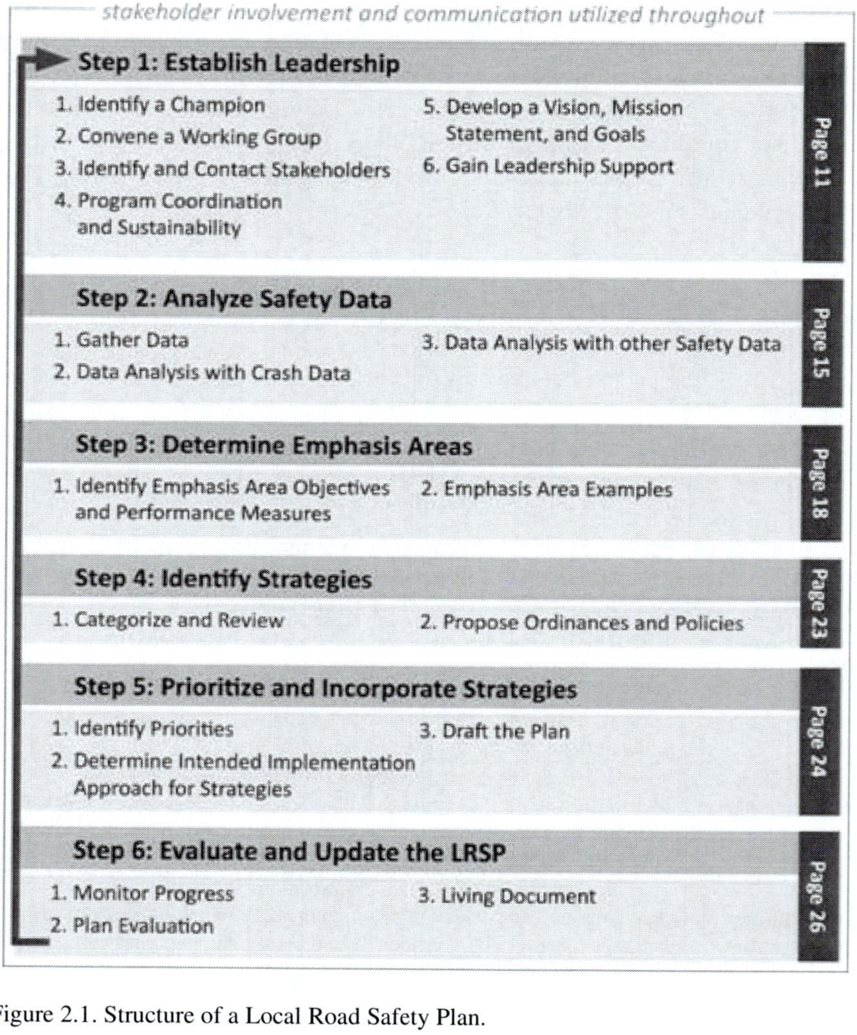


Figure 2.1. Structure of a Local Road Safety Plan.

## Critical Success Factors

The success of an LRSP is dependent on five critical components:

- Having a champion: A champion advocates for the LRSP and gathers the political support to assist in its implementation.

- Developing a clear vision and mission: A strategic vision and mission unite all stakeholders with a common goal.
- Assembling collaborative partners: Partners collaborate to implement the plan.
- Allocating appropriate resources: Manpower and management are essential for ensuring a plan's success.
- Establishing open communication: The LRSP owners should foster open and frequent communication with stakeholders, community partners, and citizens as they develop and implement the plan.

If one of these components is not initially available, the plan should still move forward, as other components may be added or expanded as the plan is refined. For example, the plan may begin with a champion, but partners may be added as the plan develops. In fact, the plan itself may help attract partners.

Practitioners who are interested in developing and implementing an LRSP should also be fully aware of the other strategic goals within their organization and how the proposed LRSP will complement those goals.

# 2. Understanding the Process of Developing a Local Road Safety Plan

As shown in Figure 2.1, developing an LRSP consists of a general six-step process. The process is cyclical; when the last step of the plan is completed the process starts all over again. The intent of this approach is to provide a framework of the key steps in developing an LRSP, but this can be tailored to the unique needs and circumstances of the locality.

## Step 1: Establish Leadership

### *Identify a Champion*

The safety champion, whether appointed or selected by the plan proponents, will advocate for the successful development, implementation, and evaluation of the LRSP. This person typically will have a keen understanding of the importance of moving forward with plan implementation and have some influence in acquiring and the use of safety resources. Champions may include a public works official, local engineer or transportation official, law enforcement officer, elected official, community administrator, or local citizen.

### *Convene a Working Group*

An LRSP can often be initiated through an informal meeting. The participants may eventually form the foundation of the LRSP working group, which is the team responsible for developing the LRSP. Bringing the right agencies or individuals together to be part of the working group will help foster a long-term commitment and build momentum to implement the plan. The responsibilities of this group include the following:

- Establishing a charter or memorandum of understanding to clarify each working group member's role.
- Analyzing data (crash, traffic, etc.) to look for trends or potential problem areas.
- Recommending and prioritizing emphasis areas to include in the LRSP;
- Engaging relevant safety stakeholders.
- Identifying public, private, and non-profit funding sources to implement the LRSP.
- Writing the LRSP.
- Marketing the LRSP through a communication plan with key messages for active public involvement.
- Encouraging local groups (civic organizations or business improvement districts) to adopt common safety goals as part of their plans.
- Participating in LRSP implementation efforts and tracking progress after the initial plan is developed.

One strategy to initiate and formalize the function and roles of the working group is to hold a kick-off meeting for the LRSP. A block of time, approximately one hour, should be set aside for this initial meeting which may be useful inidentifying stakeholders, discussing ideas, or planning a future schedule for implementation.

A sample agenda for a working group kickoff meeting is provided in Appendix A. The sample agenda includes introductions, defining the LRSP and describing the benefits, the identification of other agencies or participants that should be invited, and information for the next meeting. The specific topics included in the kickoff meeting agenda should be at the discretion of the organizer. If an initial crash analysis has been conducted, this would be appropriate to present at the project kick-off to identify the scope of the safety challenge within the locality. If possible stakeholders may begin to outline key emphasis areas; however, covering essential items such as the logistics for regular meetings may be sufficient for the kick-off meeting. The most important outcome of the kick- off meeting is to establish a commitment to create the plan and set a meeting schedule for the future.

***Identify and Contact Stakeholders***

The working group identifies and contacts LRSP stakeholders—individuals who have a vested interest in road safety. Stakeholders should include decision makers who can further the LRSP process by helping to plan, implement, and evaluate the progress of achieving the safety goals outlined in the LRSP. Typical LRSP stakeholders are listed in Table 2.1.

***Program Coordination and Sustainability***

Stakeholders working together as a team can eliminate redundancy, increase program efficiency, and leverage limited resources. Active communication among organizations, whether at the Federal, State, regional, local, or Tribal level, can create a collaborative environment that results in reduced crashes by allowing diverse agencies to share expertise and unique ideas. It is critical that working group members are identified and make a commitment to pursue the safety vision and goals. As the plan evolves, local citizens should also be consulted for input and feedback. NCHRP Report 501 offers ideas for integrating and coordinating multi-disciplinary (4E) efforts within a jurisdiction and determining the most effective combination of strategies.[6] While each safety management team is different, common themes of coordination, communication, and collaboration are paramount and are the hallmarks of a successful team approach. Regular meetings, more frequent in the development stages, should be convened to show incremental progress in the plan; this will help maintain momentum and focus. Consistent, effective communication is necessary to disseminate key information to team members and to relay key messages to the community. Finally, a collaborative environment between the involved stakeholders is required to identify barriers and to develop consensus on which safety challenges to address.

Even within a single organization, unique knowledge of assets may be known by different members of the staff or departments. An LRSP provides a framework for sharing and preserving this institutional knowledge.

**Table 2.1. Local Road Safety Plan Stakeholders**

| AREA | POTENTIAL STAKEHOLDERS |
|---|---|
| **Engineering** | • County Engineer, Road Superintendent<br>• Director of Public Works<br>• Transportation Planner / Engineer<br>• Local Engineering or Public Works Department<br>• County Highway Department<br>• State DOT Region or District Office<br>• Federal Highway Administration Division Office<br>• Bureau of Indian Affairs |
| **Enforcement** | • Chief of Police<br>• Local/Tribal Police Department<br>• County Sheriff's Department<br>• State Police/Patrol |
| **Education** | • Public Safety Stakeholders (e.g., AAA, SafeKids,<br>• Operation Lifesaver, League of American Bicyclists, etc.)<br>• Driving Education and Training Professionals<br>• Local Public Information Official |
| **Emergency Services** | • Local Emergency Service Providers<br>• Emergency Service Director<br>• Hospitals<br>• Fire Chief |
| **Other Stakeholders** | • School District (facilities and/or transportation)<br>• Local Planning and Zoning Department or Commission<br>• Governor's or State's Highway Safety Office<br>• Local Agency Budgeting Office<br>• Metropolitan Planning Organization (MPO) /<br>• Rural Planning Organization / Regional Council of Governments<br>• AARP<br>• Local Politicians/Commissioners<br>• Local/Tribal Technical Assistance Program<br>• Railroads<br>• Park District/Authority<br>• Transit Agency<br>• Community Groups (Chamber of Commerce, Tourism Agencies, etc.) |

### *Develop a Vision, Mission Statement, and Goals*

Having a clear vision is important when beginning to develop an LRSP. A vision is a description of the desired outcome of the LRSP. This may be simple, such as, "to improve road safety within our jurisdiction in order to significantly reduce the number of people being killed and seriously injured." A mission statement supports the overall vision and should provide direction. For example, the 2010 Delaware SHSP lists the following mission statement:[7]

> "Toward Zero Deaths aims to eliminate fatalities on Delaware's roadways through a multiagency approach that utilizes education, enforcement, engineering, and emergency service strategies."

Goals are set to achieve a mission and vision. Goals should be linked to the mission statement and should be realistic. For example, "saving lives and preventing serious injuries over the next decade on our local county roads" is concise and easily understood. It creates the need to move forward because it prompts action.

***Gain Leadership Support***

An LRSP can be a grassroots effort. Several LRSPs began as a basic conversation on road safety within a local agency. Gaining support from community leaders, such as a city or county council, will aid in the development of the plan and impact outcomes.

Sharing ideas and gaining feedback from a group of leaders can help gain support. People who are consulted in developing a plan will feel a greater sense of ownership, and will be stronger advocates for a plan they helped develop.

Around the country, a variety of agencies have taken the lead in the development of an LRSP. These range from city or county road agencies to an MPO or a Tribe. In States with limited local roadway ownership, the State DOT may take the lead. Several case studies illustrating how agencies have developed an LRSP can be found in Section 4.

## Step 2: Analyze Safety Data

Local road practitioners should analyze safety data prior to identifying the problem and emphasis areas. The State SHSP is a great starting point for communities considering the development of an LRSP.[8] In most cases, the SHSP has identified, through data analysis, issues on local roads in rural areas. The SHSP may have resources, data, and other information that can be used to develop an LRSP. Each SHSP typically lists the office or person responsible for the plan. In addition, there are several analysis resources available and may be in use at the State level. These resources include:

**Crash Analysis Resources** – These are guides that detail procedures for analyzing crash data. These resources include:

- Roadway Safety Information Analysis: A Manual for Local Rural Road Owners, located at http://safety.fhwa.dot.gov/local_ rural/ training/ fhwasaxx1210/
- NCHRP Report 500, Volume 21: Safety Data and Analysis in Developing Emphasis Areas, which can be found at http://www.trb.org/Publications/Public/Blurbs/Safety_Data_and_Analysis_ in_Developing_Emphasis_Ar_160164.aspx

The purpose of the latter is to provide guidance on the sources of safety data needed and on the procedures for selecting strategies within a given emphasis area and targeting those treatment strategies to either roadway locations or road-user groups.

**Highway Safety Manual (HSM)** - The first edition of the HSM provides the best factual information and tools in a useful form to facilitate roadway planning, design, operations, and maintenance decisions based on precise consideration of their safety consequences. The primary focus of the HSM is the introduction and development of analytical tools for predicting the impact of transportation project and program decisions on road safety. Further information may be found at http://www.highwaysafetymanual.org/Pages/ default.aspx.

**Interactive Highway Safety Design Model (IHSDM)** - includes six evaluation modules: Crash Prediction, Policy Review, Design Consistency, Traffic Analysis, Driver/Vehicle, and Intersection Review. The Crash Prediction Module (CPM) implements Part C (Predictive Method) of the HSM for evaluating rural 2-lane highways, rural multilane highways and urban/suburban arterials. Further information may be found at http://www.fhwa.dot.gov/research/tfhrc/projects/safety/comprehensive/ihsdm/index.cfm.

**Crash Modification** Factors (CMF) Clearinghouse - includes a Web-based database of CMFs along with supporting documentation to help transportation engineers identify the most appropriate countermeasure for their safety needs. Further information may be found at http://www.cmfclearinghouse.org/.

**SafetyAnalyst** - incorporates state-of-the-art safety management approaches into computerized analytical tools for guiding the decision-making process to identify safety improvement needs and develop a system wide program of site-specific improvement projects. Further information may be found at http://www.safetyanalyst.org/.

### *Gather Data*

Crash and other safety data are used to identify safety issues, select appropriate countermeasures, and evaluate performance. The analysis used in the development of an LRSP usually looks at the bigger picture and does not focus on analyzing crash data for a specific site. If crash data are not readily available, other safety-related data or crash risk assessments (such as locations with geometric similarities as locations of known concern) can help identify safety issues and concerns. Other data may include traffic citations, hospital records, insurance claims, speeds, traffic counts, and in some cases anecdotal evidence from safety partners.

The following additional data sources may be helpful in the development of an LRSP:

- Local law enforcement records.
- State/local crash reporting databases.
- State crash facts report.

If local agencies encounter data challenges, the *Road Safety Information Analysis: A Local Rural Road Owner's Manual* provides strategies in the collection and analysis of crash and other roadway data[9].

### *Data Analysis with Crash Data*

Crash data are the most useful to identify safety issues, however, typically at least three years of crash data are needed. Basic crash data analyses may include identifying trends based on time-based indicators (e.g., time-of-day, day-of- week, or month-of-year), environmental conditions (e.g., weather or lighting), or geographic factors (e.g., location). Examples of analyses include:

- Crashes by severity and type for the entire city/county.
- Crashes by severity and type by roadway functional classification or type.
- Contributing factors:
    - Restraint use (seat belts, car seats).
    - Alcohol or drug use.
    - Weather conditions.
    - Distracted driving.

The analyses of these data could range from simply marking locations on a map to tallying common factors in a list, or using a spreadsheet application to determine trends by location, crash type, or other contributing factors.

***Data Analysis with Other Safety Data***

When State or local-jurisdiction crash data are not available or adequate, other sources to identify safety issues include hospital/emergency responder records, towing company records, insurance databases, or by conducting road safety ratings. Road safety ratings may be based on the presence of specific roadway or roadside designs, traffic control features, or other features that can be used to assess crash risk. For example, the United States Road Assessment Program (usRAP) provides a method to identify major safety shortcomings through a program of systematic assessment of risk. The result is a star rating of roadway safety. Details on the usRAP process and methodology can be found at www.usrap.us.

An agency may conduct an observational study or a road safety audit (RSA) to gain a better understanding of safety issues. An RSA is the formal safety performance examination of an existing or future road or intersection by an independent, multidisciplinary team.[10] RSA's may also be conducted to identify safety concerns at either a single location or along a corridor. Details on RSAs can be found at http://safety.fhwa.dot.gov/rsa/.

## Step 3: Determine Emphasis Areas

The working group should identify the key emphasis areas of the LRSP. An emphasis area is an area of opportunity to improve safety through a comprehensive 4 E approach (engineering, enforcement, education, and emergency services), if appropriate. **The emphasis areas should be consistent with trends identified during the data analysis**. In some cases, if data are unavailable, emphasis areas may address concerns of the various stakeholders and the community. Local citizens should be given the chance to identify areas of concern. Methods to reach out to citizens may include public forums, open-house meetings, an internet survey, or via a request for comments advertised in the local newspaper.

Emphasis areas are an opportunity to improve safety and should reflect the input of the group and consider strategies from the 4 E's of roadway safety improvement when appropriate:[11]

- Education gives drivers information about making good choices, such as not texting while driving, avoiding alcohol or medications affecting level of consciousness, wearing a seatbelt, or informing people about the rules of the road.
- Enforcement of traffic laws and a visible police presence tend to deter motorists from unsafe driving behavior.
- Engineering addresses roadway infrastructure improvements to prevent crashes or reduce the severity of collision when they occur.
- Emergency services provide rapid response and quality of care when responding to collisions causing injury by stabilizing victims and transporting them to other facilities.

Combining the efforts of multiple strategies, such as education and enforcement can increase the likelihood of success in improving safety. The emphasis areas chosen for inclusion in the plan should reflect a balance of local issues identified by the working group and the resources (financial, expertise, and time) available to put them into practice. If an emphasis area is critical but does not currently have resources, it should still be included with the constraints listed. Sample emphasis areas are presented in Appendix B.

### *Identify Emphasis Area Objectives and Performance Measures*

Each emphasis area may help meet the plan's overall goal by establishing objectives and performance measures. Performance measures are shorter- term outcomes that contribute to achieving the strategic plan. They provide milestones, indications of progress, and should be established within a specific time period. Performance measures may be set at specified time intervals measured over the life of the plan such as, "reducing roadway departure fatalities each year and an overall reduction of 10 percent within five years."[12]

<table>
<tr><th colspan="3">Emphasis Area</th><th colspan="4">Strategic Linkage</th></tr>
<tr><td colspan="3">Intersection Safety</td><td colspan="4">Intersection safety was identified in the state-wide Strategic Highway Safety Plan as one of seven emphasis areas for the State</td></tr>
<tr><th colspan="3">Objectives</th><th colspan="4">Success Indicators</th></tr>
<tr><td colspan="3">Reduce the frequency and severity of crashes at signalized and unsignalized intersections</td><td colspan="4">A reduction in intersection crashes, particularly severe intersection crashes for the Town of Sylvia and in the surrounding areas.</td></tr>
<tr><th></th><th>Actions</th><th>Target Output</th><th>Organizations and Persons Responsible</th><th>Date of Completion</th><th>Performance Measures</th><th>Monitoring and Evaluation</th></tr>
<tr><td>Education</td><td>Public service announcements regarding dangers of red light running and stop sign violations</td><td>Awareness of the dangers of running red lights and stop signs</td><td>Ms. Naomi Fay with the Gazette is coordinating PSAs in paper and on radio station (WKAE)</td><td>Dec. 2012</td><td>Number of PSAs</td><td>Informal survey of public response planned for June public meeting</td></tr>
<tr><td>Enforcement</td><td>Enforcement blitz for high-crash intersections</td><td>Reduction in signal and stop sign violations</td><td>Chief W. McGee is organizing both blitzes and coordinating with Ms. Fay for media coverage</td><td>May and Sept. 2012</td><td>Number of tickets issued</td><td>Crashes in 2012 where red light running was cited compared to 2011</td></tr>
<tr><td>Engineering</td><td>Increase visibility by removing vegetation at intersections; place stop ahead pavement markings</td><td>Increased compliance of traffic control because of increased visibility</td><td>Mr. Haley with the county maintenance staff</td><td>Aug. 2012</td><td>Number of intersections improved</td><td>Number of intersection crashes in 2012 where sight distance was cited compared to 2011</td></tr>
<tr><td>EMS</td><td>Install emergency signal outside ambulance depot</td><td>Increased response time to intersection crashes</td><td>Mr. Luca Burton from County Public Works</td><td>Sept. 2012</td><td>Ambulance response time</td><td>Compare average response times to 2011 times</td></tr>
</table>

Figure 2.2. Emphasis Area Table.

Depending on the amount of time and number of people involved, dividing the emphasis areas among people or groups may be a useful allocation of manpower. If a person or group indicates a strong interest in targeting a particular emphasis area, they may form an emphasis area team. Emphasis area teams would be responsible for tasks such as the data analysis, strategy selection, implementation, tracking, reporting and evaluation of their assigned emphasis area.

One way to present the emphasis areas is through the use of a table that details responsibilities for implementing an action, desired outcomes, dates, and performance measure(s) for monitoring and evaluation, as shown in the example outlined in Figure 2.2. This table is also included as a template in Appendix C.

***Emphasis Area Examples***

Emphasis areas are based on the data analysis completed early in the LRSP process. Some examples include pedestrians, intersections, roadway departure, impaired driving, distracted driving, aggressive driving, commercial motor vehicles, motorcycles, and improving data.

For other examples of emphasis areas, consider reviewing the American Association of State Highway and Transportation Officials (AASHTO) Strategic Highway Safety Plan[13], which focuses on 22 key emphasis areas and contains strategies designed to improve each area's major problem areas or to advance effective practices by means that are both cost-effective and acceptable to a significant majority of Americans. The AASHTO SHSP divides the 22 key emphasis areas into six major categories: Drivers, Special Users, Vehicles, Highways, Emergency Services, and Management.

The National Highway Traffic Safety Administration's (NHTSA's) *Countermeasures that Work* also includes a variety of behavioral-related strategies that address specific emphasis areas that may be included within an LRSP.[14] Strategies address the following emphasis areas:

- Distracted and/or drowsy driving
- Occupant protection
- Impaired driving
- Speed management
- Teen drivers

- Hazardous locations
- Roadway/lane departures
- Intersection safety
- Non-motorized road users
- Older drivers
- Incident response
- Nighttime crashes
- Crash data

Appendix B provides examples of additional emphasis areas, possible performance measures, and potential strategies. Not all of the potential strategies may be applicable to all locations and their effectiveness, if applied, may vary.

## Step 4: Identify Strategies

### *Categorize and Review*

An LRSP should include a list of strategies focused on addressing the emphasis areas. Strategies will be based on identifying, categorizing, and reviewing high- priority corridors or intersections for improvement. These are locations where safety improvements are most needed to achieve the goals in the LRSP and can form the basis for system-wide improvement strategies. Strategy selection will also be based on effective and validated practices.

For example, if reducing excessive speed is an emphasis area, consider the following measures:

- The *education* component may include campaigns to educate the public about the dangers of speeding.
- The *enforcement* component may involve the use of automated enforcement.
- The *engineering* component may call for the installation of speed feedback signs.
- The *emergency service* component may require the enactment of strategies targeted at reducing response time for first responders, thus increasing the chance of survival for a person involved in a crash.

A variety of resources are available for selecting effective and validated countermeasures, including:

- Crash Modification Factor (CMF) Clearinghouse.
- FHWA Proven Countermeasures.
- Highway Safety Manual.
- NCHRP 500 Series.
- NHTSA *Countermeasures that Work.*
- Highway Safety Improvement Program (HSIP) Manual.
- Roadway Safety Noteworthy Practices Database.
- *Intersection Safety: A Manual for Local Rural Road Owners.*
- *Roadway Departure Safety: A Manual for Local Rural Road Owners.*

Please see the resources section for links to these references.

### *Propose Ordinances and Policies*

Local ordinances and local government policies addressing local road safety may help support efforts to meet the goals and objectives of the LRSP and increase public awareness of driver behavior issues. For example, if speeding is identified as an emphasis area, certain agencies may be able to pass local ordinances targeting speeding through increased fines or the use of automated speed enforcement. A champion sitting on the local council may be helpful in identifying additional policies and passing appropriate ordinances. This is often a good time to initiate an educational campaign to increase public awareness of the ordinances.

Formalizing policies can also help to improve and institutionalize safety. For example, a locality can develop polices related to maintenance of signs and pavement markings, provision of pedestrian features, transverse rumble strips, or vegetation removal. These policies can also serve as proactive risk management tools if they improve and institutionalize safety, by showing a measured approach towards improving safety.

## Step 5: Prioritize and Incorporate Strategies

### *Identify Priorities*

The proposed strategies for each key emphasis area should be prioritized by comparing the benefits and costs of implementation. This comparison can help the implementation phase by starting with the strategies that provide the highest benefit (e.g., reduction in crashes) for the least cost. However, costs and benefits are not the only considerations. Other considerations for prioritization include the availability of manpower (e.g., does the county maintenance staff have time available to trim vegetation over the summer), the schedule for implementation (e.g., are there short-term strategies that can be implemented rather quickly), and the relative importance of each emphasis area. The working group can determine an agreed upon priority for the strategy with these considerations in mind.

The Lafayette MPO case study, included in Section 4, includes three of the MPO's highest-priority challenges. These were used to define the four main goals of the plan, and the final recommendations proposed can be traced back to the priorities.

***Determine Intended Implementation Approach for Strategies***

Before drafting the plan, some thought should be given to the various approaches that will be used to implement the strategies. Some strategies will be implemented as part of a systematic improvement process such as providing rumble strips along rural corridors as part of a summer paving program. Other strategies may be part of a one-time event, such as an enforcement blitz, or a spot improvement program, such as installing protected left turn phasing at critical intersections. Because the plan will involve multiple agencies and different types of strategies, several approaches for implementation will be used.

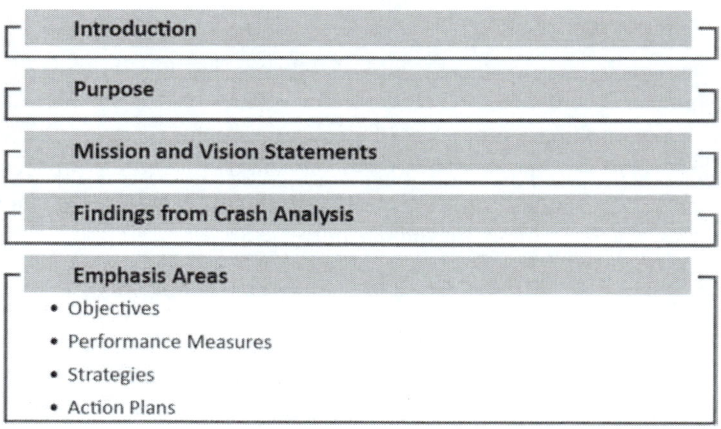


Figure 2.3. Sample Outline of Draft LRSP.

***Draft the Plan***

Depending upon the resources available, a draft plan can be created to cover the basic elements, as shown in Figure 2.3, or to create a more advanced framework for anticipated growth. An advanced framework may include emphasis areas with resource or institutional barriers or may require assistance from another level of government (e.g., adopting a primary seatbelt law).

After a brief introduction that discusses the current condition of the local jurisdiction with respect to roadway fatalities, the purpose of the plan and how the plan will help reduce fatalities and serious injuries, the plan should state the mission and vision for safety partners to work towards. The plan should discuss what data were gathered and analyzed and present the decision process regarding which emphasis areas were selected to include in the plan. Each emphasis area should identify an objective, goal, performance measures, and strategies. Defining the objective should tie back to the overall vision statement (e.g., to reduce roadway departure fatalities by 10 percent by 2020). The performance measures should monitor the progress in attaining the objective. The strategies should be listed by priority actions that can be performed by the 4 E's as appropriate. Each strategy should include a performance measure. These strategies will be incorporated in the action plans that are created for each emphasis area, as suggested in Figure 2.2.

Each section should be expanded with supporting information, such as tables or charts describing the crash data analysis results, maps detailing the area included within the plan, and photos representing areas of concern. A template for an LRSP can be found in Appendix D.

## Step 6: Evaluate and Update the LRSP

***Monitor Progress***

Successful LRSPs are monitored for implementation progress. This helps provide accountability and can be used to keep stakeholders informed and engaged. Milestones should be set to measure progress, which may entail meeting periodically to determine if strategies that support emphasis areas are being implemented. By monitoring progress opportunities for collaborating on implementing strategies could be realized, which may assist with implementation. It is also important to keep a record of the implementations to serve as a historical record of completed strategies. This data will be essential in scheduling evaluations of the strategies implemented.

### *Plan Evaluation*

Evaluation of the LRSP strategies should be ongoing to ensure the effectiveness of the projects and the overall plan. After strategies have been in place for at least one year (several years may be necessary for sufficient data), an agency may wish to evaluate their effectiveness for larger-scale implementations. In many cases a before-and-after study using crash data will not be feasible because of the unavailability or lack of crash data. When sufficient crash data are not available, other measures of effectiveness (MOEs) can be used to evaluate the safety performance of an implemented strategy. Some example MOEs include the following:

- Number and type of public comments and concerns.
- Numbers and types of police citations.
- Number of fence/wall/sign impacts.

These MOEs are observed during a field study. The MOEs should be observed under similar periods and durations before and after implementation. Traffic volume data can help provide context on the results.

If sufficient crash data are available, a simple before-after study that compares the number of crashes before implementation to the number of crashes after implementation can be conducted to determine the effectiveness of implemented strategies. Typically a decrease in crashes would indicate that the treatment has successfully improved safety. However, to verify that a decrease in crashes is due to the strategies implemented and not a reduction in trips, before and after traffic data should be collected and compared between periods. The effectiveness of each strategy should be compared to the goals for each emphasis area. This is the ultimate measure if the LRSP is achieving the desired improvements in safety.

### *Living Document*

The working group should review the LRSP, examine progress, evaluate effectiveness, and, if needed, suggest changes or modifications to the plan. This ongoing evaluation of the LRSP may present opportunities to improve the plan. Advances in roadway development, legislation, and technology may also invite opportunities to update the plan. The working group should review these advances for possible incorporation into the plan. Regularly scheduled updates allow the working group to review what is working well and adjust what needs improvement. Establishing a regular evaluation and update cycle can assure routine examination of the plan and maximize the plan's effectiveness.

# 3. Common Issues, Opportunities, and Challenges

When developing an LRSP, agencies may be confronted with challenges from lack of personnel and data to funding restrictions and/or limitations.

Resolving these issues provides an opportunity to build new alliances, collaborate with other organizations, and secure new sources of funding. Guidance related to these challenges follows.

## Personnel

Local Technical Assistance Program (LTAP) and Tribal Technical Assistance Program (TTAP) centers, universities and other research institutions that work on transportation safety issues may also be available to provide assistance in the development of an LRSP. State DOTs and Regional MPOs or Regional Planning District Commissions may be able to assist with an LRSP. These agencies have resources or information on the development of LRSPs and about other LRSPs previously developed.

Some agencies have engaged a consultant to help create their LRSPs. A consultant may be able to bring additional expertise to the project.

Other organizations, such as AAA and local community groups, may be able to assist in the development of an LRSP as well. This may include creating a broad coalition of participants such that volunteers and other resources can be leveraged.

## Funding

Limited funding is often a concern to most road agencies as they seek to implement roadway safety improvements. Agencies may also be concerned about how they will fund projects identified within an LRSP. In many cases, having an LRSP in place will increase an agency's ability to secure funding to implement its safety strategies when competing for funding resources. An LRSP can also be useful for leveraging private funds from local businesses and corporations interested in investing in traffic safety. An LRSP can be developed using in-house staff in coordination with other agencies, thereby limiting the costs of time and financial resources spent developing the plan.

Several options for funding both the development of an LRSP and the implementation of its strategies are discussed below.

- The State and Community Highway Safety Grant Program (Section 402) was developed to provide funding to implement initiatives targeted at improving safety. Funds are typically used for safety projects related to enforcement, education, and emergency services and are administered by each State's DOT or highway safety office. Additional information is available at http://safety.fhwa.dot.gov/policy/section402/.
- Federal funds within the Highway Safety Improvement Program (HSIP) may be used to develop an LRSP, as well as be used to implement the infrastructure-based improvements identified within an LRSP. HSIP funding is administered by each State's DOT; the process to apply for funding varies by State. More information about HSIP can be found at http://safety.fhwa.dot.gov/hsip/.
- Contacting the State's DOT or other county and local transportation entities may help identify other funding opportunities. An MPO or Regional Planning Organization (RPO) may be able to allocate financial assistance for developing a plan. While money may appear to be a significant hurdle to the implementation of an LRSP, funding sources are available, and having an LRSP will typically increase an agency's ability to secure funding to implement its strategies.

## Limited Data

Limited data should not inhibit the development of an LRSP. Sources of crash data include NHTSA's Fatality Analysis Reporting System (FARS), the State DOT, raw law enforcement data, or the existing SHSP.

Data, such as traffic citations, hospital records, and speed monitoring, can also be used in the development of an LRSP. Other risk assessment processes, such as road safety audits or the usRAP, which provide methods to identify major safety shortcomings through a program of systematic assessment of risk as discussed in Section 2, can be used to identify safety issues for an LRSP.

In many cases, agencies use their LRSPs to develop strategies to improve their local data. By including data collection as an emphasis area within an LRSP, an agency or group of agencies can begin to assess gaps in data and develop strategies for improvement. *Roadway Safety Information Analysis: A Manual for Local Rural Road Owners* was developed by FHWA to provide strategies in the collection and analysis of crash and other roadway data. This document can be found at http://safety.fhwa.dot.gov/local_rural/training/fhwasaxx1210/.

Existing State and regional databases may also be used as a model for developing and implementing a data collection program. Data collected may include road conditions, maintenance records, weather, or other environmental factors, or any other data that may help to generate a snapshot of safety concerns. To find out if the State DOT may be able to provide this information, refer to http://www.dottrcc.gov.

# 4. Case Studies

LRSPs have been implemented by a variety of jurisdictions around the US. This section contains examples of programs implemented to address local rural safety issues.

## Ongoing Commitments to Road Safety

### *Champaign County, Illinois*

The agencies in Champaign County mapped out a unified strategy in a multi- disciplinary LRSP (originally called a Comprehensive Highway Safety Plan). This was an effort to continue their commitment to road safety following the State Farm Embedded Safety Specialist Pilot Initiative (ESSI) which was launched in December 2006 in Champaign County, Illinois. ESSI was designed to develop and implement strategies for improving safety by reducing crashes instilling a more integrated safety culture within participating agencies.

The plan envisioned cooperation by various agencies. Each agency's engineering department would be able to accomplish the engineering aspects of the plan, having both the authority and responsibility to build and maintain a safe road system. Challenges were also identified, including securing sustainable funding and developing partnerships with organizations that can assist with the education and enforcement aspects of the plan.

The plan focused on several key objectives:

1. Recommending proven safety measures.
2. Providing a structured and realistic set of responses that implement changes over time.

3. Integrating a 4-E approach in its proposed solution strategies.
4. Identifying road safety partners that could sustain a long-term effort.

The process for developing the plan included:

- **Identification of the Champion and Safety Partners:** The champion was the MPO, Champaign-Urbana Urbanized Area Transportation Study (CUUATS). CUUATS took the lead on the implementing the process. An important part of the plan process was the identification of road safety partners already in the community. The LRSP served as a basis for setting up a county-wide safety committee which included engineering, enforcement, education, and emergency service representatives. The partners provided substantial input associated with problem identification and prioritization, as well as recommended solutions that were incorporated into the Comprehensive Highway Safety Plan.
- **A Gap Analysis:** As a part of the Comprehensive HighwaySafety Plan process, agency interviews were conducted to identify safety concerns.
- **Network Screening:** Network screening was a systematic process that extracted from the Illinois Department of Transportation (IDOT) Local Accident Reference System (LARS) database useful information on the risk and potential mitigation of crashes at locations in Champaign County.

The following 13 key road safety issues were identified during the development process:

1. Intersections.
2. Information for decision-making.
3. Alcohol and other impaired drivers.
4. Driver behavior and awareness.
5. Roadway departure.
6. Vulnerable users.
7. Traffic signs.
8. Highway-railroad grade crossings.
9. Work zones.
10. Large trucks.
11. Safety belts / occupant protection.

12. Congestion.
13. Travel modes.

A set of 36 education, enforcement, engineering, emergency service, and institutional strategies was developed to improve safety in response to these concerns. These strategies were summarized and timeframes were recommended for implementation. Through the committee, a variety of intersection and corridor studies were implemented within the county leading to a number of RSA's, HSIP-funded safety projects, and a teen driver safety initiative.

## Obtaining Data and Establishing Relationships are High Priorities

***Lafayette MPO, Lafayette Parish, Louisiana***

Financed through grants from FHWA and the Federal Transit Administration (FTA), the Lafayette Consolidated Government MPO sought to develop an LRSP (originally called a Transportation Strategic Safety Plan).

The plan included more than 250 strategies that incorporated the 4 E's and varied by cost, proven effectiveness, and the number of partners, among other factors. Lafayette Parish has a high fatality crash rate that is well above the State average.

High-priority challenges were identified: obtaining crash data from parishes and municipalities included in the MPO study area, re-establishing and improving current relationships among enforcement agencies and emergency services, and establishing new relationships among enforcement agencies and Emergency services in parishes now included in the MPO Study Area.

More than 30 agencies were identified as key stakeholders in fostering collaboration, including school systems, law enforcement agencies, cities, departments of safety and transportation, and citizen committees.

Four main goals were proposed in the plan:

1. Protection of drivers and passengers.
2. Protection of all roadway users, such as pedestrians, bicyclists, and motorcyclists.
3. Improvement of infrastructure.
4. Implementation of system-wide programs.

The most important goal was driver and passenger protection. Louisiana has been a "primary" enforcement State (i.e., one in which a driver may be pulled over solely for not wearing a seatbelt without needing any other visible infraction) with regards to safety belt usage since 1995, as the most obvious form of occupant protection is the safety belt.

To achieve these goals, the plan proposed the following actions:

1. Maximizing the use of occupant restraints by all vehicle occupants.
2. Insuring that restraints, especially child and infant child safety seats, are used properly.
3. Providing access to appropriate information, materials, and guidelines for implementing programs to increase the use of safety restraints among vehicle occupants.

The plan also listed all of the objectives and strategies along with the associated lead department or agency responsible for their implementation to include partners, approximate cost, and effectiveness.

Two primary measures of effectiveness for the implementation of these strategies were identified:

1. Safety belt use rate and percentage for the Lafayette MPO study area.
2. Number of unbelted fatal and serious injury crash victims and proportion of all fatalities.

The plan includes details about annually reviewing performance measures and guidelines as to how changes may be implemented. Additionally, there is a section devoted to funding implementation and continued management of the plan.

For the 2011-2012 fiscal year, Lafayette MPO proposed the following five recommended actions in their implementation plan totaling $3,210,000:

- Know Your Traffic Law Program - To educate the public about traffic safety and other risky behavior commonly experienced by Lafayette motorists.
- Safety Town – To construct a permanent, interactive education facility where children learn about traffic safety in a safe environment.
- Safety Fair – To educate youth about the basics of road safety.

- DWI Crash Emergency Room Program – To provide mobile trauma room demonstrations from high school to high school.
- Judicial Community Service Plan for Traffic Violators – To have judges assign community service hours based on the MPO Safety Plan.

## LRSP Generated Using SHSP Model

### *Olmsted County, Minnesota*

Olmsted County prepared an LRSP as part of an effort to reduce the number of fatal and serious injury crashes that occurred on its system of highways. It was part of a coordinated effort with the Southeast Minnesota Towards Zero Deaths organization and the sheriff's department to identify opportunities for targeted enforcement campaigns.

The traffic safety priorities identified were based on the data-driven analysis of approximately 11,000 crashes that occurred in Olmsted County between 2003 and 2007. The analysis was done in accordance with the guidelines for Federal transportation safety funds and Minnesota's SHSP.

These primary steps were performed in completing the plan:

- Conducting a comprehensive crash analysis, including system (i.e., State vs. local), severity, location (i.e., urban vs. rural), intersection-related vs. road departure, and crash type.
- Identifying Olmsted County's safety emphasis areas (22 identified causes of crashes).
- Identifying a short list of high-priority proven safety strategies effective at reducing specific types of crashes.
- Identifying locations on the county's highway system that were most at-risk based on a variety of system-wide factors.
- Identifying safety improvement projects that would be eligible for funding through the State's HSIP.

Following the same procedure used to generate the State's SHSP, Olmsted County identified the following emphasis areas as being highest-priority on its local highway system:

- Driver behavior – young drivers, aggressive driving, impaired driving, and seat belt usage.
- Specialist users – bicyclists.
- Highways – road departure and intersections.

Potential safety improvement strategies for each of the emphasis areas were evaluated. The initial list came from the NCHRP 500 Series Reports. Some proposed strategies were then eliminated from further consideration because they were considered too expensive or were experimental strategies that had no history of application in Minnesota.

A safety strategies workshop was held in Olmsted County and was attended by 50 safety partners representing the State, county, townships, and cities, as well as the Mayo Clinic, bicycle advocates, and driver education professionals. The two primary objectives of the workshop included sharing the results of the analytical process and providing a forum to discuss the short list of safety strategies. The participants further prioritized the safety strategies from highest to lowest.

As a result of the strategies identified in the LRSP, safety strategies were implemented county-wide, not just at specific locations. These strategies included the implementing the following safety measures and resulted in $962,000 of federally funded improvements over three years:

- Improving the edges of rural highways and enhancing delineation of horizontal curves in rural areas.
- Upgrading the signs and pavement markings, installing street lights, and providing dynamic warning signs at rural stop-controlled intersections.
- Adding technology at signalized intersections to support increased enforcement levels for red light running.
- Adding channelization and median islands to restrict/control turning maneuvers at urban stop-controlled intersections.

The potential improvements were low-cost in nature and aligned with the State's SHSP, placing Olmsted County in a position to qualify for HSIP funding. The County LRSP also included a section on policy addressing edge line rumble strips and rumble stripes, pavement markings, and pedestrian crosswalks.

## Securing Funding and Developing Partnerships

### *Sokaogon Chippewa Band of Mole Lake, Forest County, Wisconsin*

The Wisconsin Department of Transportation (WisDOT) commissioned a multi- disciplinary Tribal Highway Safety Plan to map out a unified strategy to address highway safety on tribal lands within Mole Lake, Wisconsin.

Several challenges associated with implementing the plan were identified, including securing sustainable funding and developing partnerships with organizations that can assist with the education and enforcement initiatives outlined in the plan. Key stakeholders were identified and partnerships were formed between the Sokaogon Chippewa of Mole Lake, WisDOT, the Forest County Sheriff's Department, and the Forest County Highway Administration.

The following represent the five key emphasis areas identified in the plan:

1. Interaction of Modes (All-Terrain Vehicles [ATVs], Snowmobiles, Dirt Bikes).
2. Speeding.
3. Pedestrians and Bicycles.
4. Large Trucks.
5. Enforcement.

Objectives were identified for each of the key safety issues, with short-, medium-, and long-term initiatives in engineering, enforcement, education, and emergency services proposed for each.

Speeding was identified as a great concern within Mole Lake, and targeted in the following safety objective:

> *Reduce speed- and behavior-related collision frequency and severity, and promote observance of speed limits and traffic control devices to improve safety for all road users.*

Potential strategies were identified to reduce speeding in the areas of engineering, enforcement, and education.

**Education**

- Targeted education campaigns to high-risk populations (short- and medium-term).
- Public Service Announcements (short- and medium-term).

**Enforcement**

- Coordination with WisDOT on speed enforcement grant through the Bureau of Transportation Safety (BOTS) or the Bureau of Indian Affairs (BIA) (short-term).
- Targeted enforcement, by time or location (short- and medium-term).
- Use of permanent speed feedback displays at locations where speeding is a safety concern (short-term).

**Engineering**

- Install gateway treatments on the northbound and southbound approach to the Mole Lake business district (long-tem).
- Install a two-way left turn lane through Mole Lake as a strategy to slow driver speeds (medium-term).

Regarding non-motorized safety, there was a significant pedestrian safety issue on a section of the state highway that crossed Swamp Creek. As a result of the LRSP, an RSA was conducted and the Tribe was able to secure American Recovery and Reinvestment Act funding (ARRA) to build a bridge for non-motorized users parallel to the state highway. In addition, sidewalk and marked pedestrian crossings were installed throughout the reservation.

The enforcement emphasis area dealt with partnerships. The Forest County Sheriff was contracted to provide enforcement on the reservation; however, the Department only had authorization to enforce traffic laws on the state highway and county roads. Therefore, the Sheriff had no authorization to enforce traffic laws on the BIA roads and they could only respond to crashes if called by the tribe. The LRSP presented several strategies for evaluating the options to provide enforcement on the BIA roads.

## Developed a Model and Framework

### *Wyoming Rural Road Safety Program*

The Wyoming LTAP Center developed a Wyoming Rural Road Safety Program (WRRSP) with funding from Wyoming DOT (WYDOT), Mountain Plains Consortium (MPC), and FHWA, and in cooperation with Wyoming counties. The main objective of the WRRSP was to develop and evaluate a transportation safety program that could help local agencies reduce crashes and fatalities on rural roads statewide. It was written to provide other local agencies with a framework to be used if they were interested in implementing a rural road safety program.

Through this LRSP concept, local jurisdictions qualify for HSIP funding from the Wyoming DOT. Through the Spring of 2011, the WRRSP has allocated more than $1.5 million to implement low-cost safety improvements.

As part of the study, a Local Road Safety Advisory Group (LRSAG) was established, including representatives from WYDOT, the Wyoming LTAP, Wyoming Association of County Engineers and Road Supervisors (WACERS), the Wyoming Association of Municipalities (WAM), and FHWA.

Four subtasks were identified in the WRRSP:

1. Identify roadway classifications systems used by counties in Wyoming.
2. Develop a methodology for using available data (crash records, traffic volume, speed, etc.) for crash prediction on rural roads.
3. Establish a five-step methodology to identify safety measures on high-risk rural roads.
4. Develop a procedure to perform economic analysisfor safety measures.

The report consisted of seven sections:

1. Introduction.
2. Summaries of literature reviews for each of the three research objectives.
3. Introduction of the detailed procedure of the WRRSP.
4. The roadway classification survey and its results.
5. The regression model methodology used to predict crashes on rural roads.
6. The procedure for performing economic analysis for safety improvements.
7. Summation of the conclusions and recommendations for future studies.

The plan also provides forms, guidelines, photos, surveys, and sample data useful for implementing this framework in other jurisdictions.

## Conclusion

Approximately 60 percent of all road miles in the U.S. are non-Interstate, rural roads owned and operated by local entities, such as towns, counties, and tribal governments. In 2009, 56 percent of highway deaths occurred on rural roads, though only 23 percent of Americans live in rural areas. Furthermore, the fatality rate was 2.6 times higher in rural areas than in urban areas, underscoring the need to systematically improve road safety in rural areas.

All States already have a comprehensive safety plan that provides a framework for reducing highway fatalities and serious injuries on all public roads; this is known as the State's Strategic Highway Safety Plan (SHSP). The SHSP is a data- driven plan that establishes statewide goals, objectives, and key emphasis areas that integrates the 4 E's of Safety – engineering, education, enforcement and emergency services. The SHSP can assist local practitioners in addressing safety on local rural roads but a locally-focused plan is often needed to address the unique conditions that contribute to safety problems and to assist local practitioners in making informed safety investment decisions. These challenges faced by local agencies can be addressed through the creation of a Local Road Safety Plan (LRSP).

An LRSP is a locally-coordinated safety plan that provides a comprehensive framework to identify key safety needs and guide safety investment decisions on local rural roads. An LRSP provides an excellent opportunity for agencies at all levels of government (local, State, and Federal) and other stakeholders to work together to align and leverage resources and funding to address the safety challenges unique to rural roads.

In general, the LRSP development process involves six steps:

Step 1: Establish Leadership
Step 2: Analyze Safety Data
Step 3: Determine Emphasis Areas
Step 4: Identify Strategies
Step 5: Prioritize and Incorporate Strategies
Step 6: Evaluate and Update the LRSP

Successful development of an LRSP depends on fostering leadership and communication among various stakeholders. This can start with the identification of a safety champion. The safety champion advocates for the plan's successful development, implementation, and evaluation. It is also important to establish an LRSP working group. This is the team responsible for developing the LRSP, performing duties ranging from defining each working group member's role to tracking progress after the initial plan is developed. The working group also identifies and contacts other stakeholders who can further the LRSP process by helping to plan, implement, and evaluate the progress of achieving the safety goals outlined in the LRSP.

Local road practitioners should analyze any available safety data to identify problem areas that will be addressed in the LRSP. Crash data should be used to identify safety issues. Typically, three years of crash data are needed to average out those years of extreme numbers. If crash data are not readily available, other safety-related data can be used to identify safety issues. The analyses of these data could range from simply marking locations on a map to using a spreadsheet to determine trends by location, crash type, or other factors.

The working group should identify emphasis areas consistent with trends identified during the data analysis and the concerns of the various stakeholders. Strategies to address emphasis areas should consider the 4 E's to comprehensively address safety. The proposed strategies for each emphasis area should be prioritized based on benefit (e.g., reduction in crashes) verses cost, the availability of manpower, the schedule for implementation, and the relative importance of each emphasis area. Some strategies can be implemented as part of a systemic improvement process such as providing rumble strips along rural corridors as part of a summer paving program or conducting an enforcement blitz. Detailed information on effective strategy selection may be acquired through partnerships with State or local agencies or by contacting the State Local Technical Assistance Program (LTAP).

LRSPs should be monitored for progress to ensure implementation of strategies that support emphasis areas and to determine if new strategies need to be considered. This helps provide accountability and can be used to keep stakeholders informed and engaged. Evaluation of the LRSP strategies should be ongoing to ensure the effectiveness of the projects and the overall plan. Addressing safety on local rural roads can be challenging. The development of an LRSP can serve as a cornerstone to building a comprehensive safety program to address the safety challenges on the roadways. Depending on needs and jurisdiction, the LRSP will vary in size and level of detail. The

LRSP is a living document and should be revisited as established goals are achieved.

# APPENDIX A. TEMPLATE FOR KICKOFF MEETING AGENDA

**[INSERT AGENCY NAME]** Local Road Safety Plan
Kick-off Meeting

Date and Time: **[INSERT DATE AND TIME]**
Location: **[INSERT MEETING LOCATION AND ADDRESS]**

1. Welcome remarks by the Local Road Safety Plan coordinator or champion
2. Introductions
3. What is a Local Road Safety Plan – Presentation outlining what a Local Road Safety Plan is and how developing one can benefit **[INSERT AGENCY NAME]**.
4. Identify other agencies or individuals who should be invited to join the working group.
5. Summarize Data Analysis – An initial data analysis has been conducted prior to this meeting to provide the group with background information on a variety of potential safety issues in **[INSERT AGENCY NAME]**.
6. Begin Identifying Emphasis Areas for the Plan
   a. Education
   b. Enforcement
   c. Engineering
   d. Emergency Services
7. Next Meeting – Schedule a date for a follow-up meeting
8. Adjourn
9. Meeting Contact:
   **[INSERT PHONE NUMBER AND EMAIL FOR MEETING ORGANIZER]**

# APPENDIX B. SAMPLE EMPHASIS AREAS

| Emphasis Area | Measure(s) | Potential Strategies |
|---|---|---|
| Reduce Distracted Driving | • Reduce annual distracted driving crashes from 100 in 2010 to 90 or fewer in 2020 (a 10-percent reduction).<br>• Reduce annual distracted driving injuries from 40 in 2010 to fewer than 36 in 2010 (a 10-percent reduction). | • Develop a public outreach campaign that coincides with other jurisdictions' efforts to raise awareness about distracted driving.<br>• Reduce roadside distractions.<br>• Pass and enforce legislation that specifically penalizes distracted driving, including making distracted driving a subsection of negligent driving.<br>• Increase the use of techniques that limit the frequency and severity |
| Increase Occupant Protection | • Reduce annual unrestrained crashes from 20 in 2008 to 5 or fewer in 2018 (a 75-percent reduction).<br>• Reduce annual unrestrained injuries from 16 in 2008 to fewer than 4 in 2018 (a 75-percent reduction).<br>• Increase safety belt use from 75 percent in 2008 to 90 percent or greater in 2018 (a 17-percent increase). | • Develop an incentive/ recognition program for law enforcement efforts.<br>• Conduct an enforcement program that targets pickup truck drivers and passengers.<br>• Conduct sustained high-visibility enforcement initiatives.<br>• Continue current best practice enforcement and educational programs (e.g., Chiefs' Challenge, Click It or Ticket).<br>• Provide more paid media campaigns in local publications.<br>• Conduct outreach to teens. |
| | • Reduce annual alcohol-related crashes from 9 in 2009 to 3 or fewer in 2014 (a 67-percent reduction).<br>• Reduce annual impaired driving injuries from 10 in 2009 to fewer than 2 in 2014 (an 80-percent reduction). | • Increase the number and effectiveness of sobriety checkpoints and targeted enforcement.<br>• Enact stronger and more effective legislation, such as license suspension or mandatory ignition interlocks for first time offenders.<br>• Develop educational programs targeting specific audiences, such as elementary and middle school students, or the 18-to-34 year-old age group.<br>• Create effective media campaigns in both visual and print media. |

**Appendix B. (Continued)**

| Emphasis Area | Measure(s) | Potential Strategies |
|---|---|---|
| Speed Management | • Reduce annual aggressive driving injuries from 24 in 2009 to fewer than 20 in 2014 (a 17-percent reduction). | • Change the driving culture by conducting and supporting public education and outreach activities that elevate the awareness of the dangers of aggressive driving. |
| | • Reduce annual fatalities involving excessive speed from 15 in 2009 to 10 or fewer in 2014 (a 50 percent reduction).<br>• Reduce annual injuries involving excessive speed from 100 in 2009 to fewer than 75 in 2014 (a 25-percent reduction). | • Educate the judiciary and elected officials on the risks associated with aggressive driving.<br>• Communicate the factors associated with aggressive driving to the transportation engineering and planning communities.<br>• Increase enforcement targeting aggressive driving. |
| Develop Safe Teen Drivers | • Reduce annual crashes involving teen drivers from 200 in 2009 to 150 or fewer in 2020 (a 25-percent reduction).<br>• Reduce annual injuries involving teen drivers from 50 in 2009 to fewer than 25 in 2020 (a 50-percent reduction). | • Review, evaluate, and improve the driver preparation program.<br>• Develop a program to increase enforcement, prosecution, and adjudication of young driver traffic law violations.<br>• Identify opportunities for engineering solutions to prevent young driver crashes through road safety audits and other measures. |
| Eliminate High- crash Locations | | • Conduct a public information and education campaign targeting the various aspects of hazardous locations.<br>• Identify best practices and innovative enforcement techniques to eliminate high-crash locations in high-crash pedestrian locations and in work zones.<br>• Conduct road safety audits targeting high-risk pedestrian and intersection locations to determine the contributing crash factors and identify effective countermeasures.<br>• Develop and implement projects to reduce or eliminate safety hazards or otherwise to enhance safety for road users. |

| Emphasis Area | Measure(s) | Potential Strategies |
|---|---|---|
| Reduce Roadway/ Lane Departures | • Reduce annual run-off-the-road fatalities from 8 in 2009 to 4 or fewer in 2020 (a 50-percent reduction).<br>• Reduce annual run-off-the-road injuries from 40 in 2009 to fewer than 30 in 2020 (a 25-percent reduction). | • Implement stricter law enforcement of motor vehicle laws and increase fines for serious violations that result in run-off-the road crashes (e.g., driving too fast for conditions).<br>• Improve data collection and analysis for fatal and injury run-off-the-road crashes to provide critical information to transportation planners and engineers.<br>• Evaluate pavement strategies to reduce speed and increase friction (e.g., pavement type, pavement application method, pavement marking spacing).<br>• Improve traffic control strategies to provide positive guidance to keep vehicles on the road.<br>• Implement forgiving roadway designs that mitigate the impact of cars leaving the road. |
| Improve Incident Response | • Reduce crash injury mortality by 10 percent or greater by 2015. | • Improve electronic data and voice communications for emergency response.<br>• Improve resource deployment for emergency service response.<br>• Develop a safer, faster EMS response.<br>• Improve crash scene safety.<br>• Improve patient care. |
| Improve Intersection Safety | • Reduce annual intersection-related fatalities from 12 in 2009 to 6 or fewer in 2020 (a 50-percent reduction).<br>• Reduce annual intersection-related injuries from 60 in 2009 to fewer than 45 in 2020 (a 25-percent reduction). | • Reduce the number of conflict points and provide better guidance for motorists at intersections.<br>• Develop a system to track and evaluate countermeasure effectiveness at high-crash intersections.<br>• Encourage more multidisciplinary collaboration at the State and local level on intersection safety.<br>• Create intersection safety checklists for existing conditions and new design. |

## Appendix B. (Continued)

| Emphasis Area | Measure(s) | Potential Strategies |
|---|---|---|
| Increase Non-Motorized Road User Safety | • Reduce annual pedestrian injuries from 15 in 2009 to fewer than 10 in 2015 (a 33-percent reduction). | • Conduct public education and outreach to motorists to raise their awareness of pedestrian and bicyclist safety needs.<br>• Conduct periodic roadway safety assessments of locations with growing traffic and pedestrian volumes and locations at greatest risk for pedestrian fatalities and injuries, and share information with other local partners.<br>• Implement effective countermeasures for problem areas as determined by roadway safety assessments.<br>• Educate the judiciary on the importance of penalties for violation of pedestrian laws.<br>• Identify opportunities for alternate funding; for instance, the Safe Routes to School Program. |
| Increase Safe Driving Behaviors in Older Drivers | • Reduce annual fatalities involving drivers 65 years or older from 12 in 2009 to 8 or fewer in 2020 (a 33-percent reduction).<br>• Reduce annual injuries involving drivers 65 years or older from 120 in 2009 to less than 100 in 2015 (a 17-percent reduction). | • Develop effective methods to identify at-risk older drivers.<br>• Develop enhanced training for emergency service personnel on the proper assessment and triage of older persons at crash scenes.<br>• Incorporate the Federal Highway Administration's (FHWA) Older Driver and Pedestrian Guidelines into the local design guidelines. |
| Reduce Nighttime crashes | • Reduce annual nighttime fatalities from 36 in 2009 to 32 or fewer in 2015 (an 11-percent reduction).<br>• Reduce annual nighttime injuries from 96 in 2009 to fewer than 80 in 2015 (a 17-percent reduction). | • Review, evaluate, and replace signage showing wear or reduced retroreflectivity.<br>• Review and evaluate locations exhibiting higher incidence of nighttime crashes for possible supplementary lighting. |

| **Emphasis Area** | **Measure(s)** | **Potential Strategies** |
|---|---|---|
| Improve Crash Data Reporting | • Improve crash report submission time from 90 days to 30 days. | • Identify best practices in crash reporting, as well as barriers to complete, accurate, and timely information.<br>• Standardize crash reporting policies and protocols for all agencies operating within the jurisdiction.<br>• Identify tools to increase timely and accurate reporting (e.g., GPS devices and laptops in responding vehicles). |

# APPENDIX C. EMPHASIS AREA TABLE

| **EMPHASIS AREA** | | | **STRATEGIC LINKAGE** | | | |
|---|---|---|---|---|---|---|
| | | | | | | |
| **OBJECTIVES** | | | **SUCCESS INDICATORS** | | | |
| | | | | | | |
| | **Actions** | **Target Output** | **Organizations and Persons Responsible** | **Date of Completion** | **Performance Measures** | **Monitoring and Evaluation** |
| **Education** | | | | | | |
| **Enforcement** | | | | | | |
| **Engineering** | | | | | | |
| **EMS** | | | | | | |

# APPENDIX D. TEMPLATE FOR AN LRSP

## Introduction

### *Background*

Discuss the current condition of the local jurisdiction with respect to roadway fatalities. Discuss the purpose of the plan and how the plan will help reduce fatalities and serious injuries.

**Vision:** A vision statement describes what the group is striving to achieve.

**Mission:** The mission statement defines the purpose of the plan, what it does, and what it is about. The mission statement usually does not change and helps to define organizational culture.

**Goals:** What does the group hope to achieve in a specified amount of time? Goals should be realistic and measurable and can evolve over time.

## Safety Partners/Stakeholders

List the agencies that were consulted in the development of the LRSP and are crucial to achieving plan goals. The following is a list of potential safety partners. Expect to engage a variety of organizations in plan development and implementation. (List Organizations Only)

- Local engineering or public works department
- County Highway Department
- State DOT Region or District Office
- Federal Highway Administration Division Office
- Bureau of Indian Affairs
- Local/Tribal Police Department
- County Sheriff's Department
- State Police/Patrol
- Local public information official
- Governor's Highway Safety Office
- Public safety stakeholders (i.e., AAA, SafeKids, Operation Lifesaver, League of American Bicyclists, etc.)
- Local emergency service providers
- School district (facilities and/or transportation)

- Local planning and zoning department or commission
- Local agency budgeting office
- Metropolitan Planning Organization
- Consultants
- Local/Tribal Technical Assistance Program
- Railroads
- Community groups (chamber of commerce, tourism bureaus, etc.)

## Metholodology

Discuss the process used to develop the plan, recognizing that the LRSP development process is every bit as important as the actual finished product. Because there will be many activities that lead to the LRSP, it might be difficult to determine whether the required process was followed just by reading the LRSP. For this reason, it is recommended that some documentation or explanation of the process be included in the LRSP. The LRSP is a living document and is expected to evolve over time.

Turnover of team members is inevitable, and it should also be expected that many who were not directly involved in the development process will read and use the LRSP. By including some explanation of the development process, both new team members and those existing ones who were not as closely involved will be able to follow the progression of the plan's creation. A section on the development process or reference to documentation will also help approving officials.

## Data Analysis

Discuss how the data were gathered and analyzed. Include any concerns about the quality and reliability of data, improvement needs, and assumptions.

## Emphasis Areas

List emphasis areas, explain how they were determined, and why addressing them is vital to achieving plan goals. The list will serve as a quick reference and introduction to the body and structure of the plan. The following information should be provided for each emphasis area and can be entered into the chart located in Appendix C.

### *Background*

The background for the emphasis area should include a brief explanation regarding why this was important to the LRSP. In this section, the writers should provide fatality data and show trends via graphs and charts to demonstrate need to address this issue.

### *Objective*

What is the objective for this emphasis area? (E.g., to reduce roadway departure fatalities by 10 percent by 2020.)

### *Performance Measures*

What measures will be used to monitor progress in attaining this objective after implementation?

### *Strategies*

List the strategies that will be performed. These strategies should include actions that can be performed by the 4 E's, if appropriate. Each strategy should include a performance measure. These strategies will be carried over as action plans that are created for each emphasis area. Further information about what, how, when, where, and who of safety activities will be included in the action plans. Action plans can also provide specifics such as various funding sources for safety activities and may also contain some project-level detail, responsible agencies, and timeframes for safety activities.

## Implementation Process

Discuss how the group plans to implement the LRSP. Include a schedule of projects or how projects will be scheduled (i.e., action plan).

## Evaluation Process

What will the evaluation process entail and how often will evaluation take place? Is someone responsible for monitoring progress throughout the year,and when will plan revisions be made? How will the LRSP evaluation affect future projects funded through HSIP or other funding sources? How will the project evaluations in these programs affect the LRSP?

## Next Steps

To keep everyone engaged and on task, discuss what comes next and who is responsible for what and when. What are the partners' responsibilities and what will each of them do with the plan? Remember, this is a living document!

## References

This section could list other plans that were referenced in the development of the LRSP, such as the State's SHSP, as well as provide more detail on the emphasis areas described in the plan. The LRSP is a strategic planning document and is intended to be concise. Action plans are based on the emphasis areas outlined in the LRSP and expand on the supporting data and strategies. These details would describe the what, how, when, where, and who. Action plans can also provide specifics on funding and some project-level detail. They may also include evaluation criteria for assessing the success of the implemented safety strategies. Ideally, each emphasis area in the LRSP should be supplemented with an action plan.

# APPENDIX E. RESOURCES

FHWA has a number of resources on its website for the Local and Rural Road Safety Program. These include information about crash facts; funding, policy, and guidance; safety programs; and partners and resources safety.fhwa.dot.gov/local_rural.

The National Association of County Engineers (NACE) has a page on its website (www.countyengineers.org) devoted to the promotion of safety on country roads.

The Center for Excellence in Rural Safety (CERS) at the University of Minnesota maintains a Web site with news, webinars, and publications www.ruralsafety.umn.edu.

Strategic Highway Safety Plan (SHSP). safety.fhwa.dot.gov/safetealu/shspquick.cfm safety.fhwa.dot gov/hsip/shsp

National Highway Institute (NHI), Low-Cost Safety Improvements Workshop
Instructor Lead: FHWA-NHI-380076 http://www.nhi.fhwa.dot.gov/training/course_detail.aspx?num=FHWA-NHI-380076&get=all
Web Based: FHWA-NHI-380083
http://www.nhi.fhwa.dot.gov/training/course_detail.aspx?num=FHWA-NHI-380083&cat=t&num=380

ATSSA, NACE, Low-Cost Local Rural Road Safety Solutions
docs.mvrpc.org/safety/Low_Cost_Local_Roadway_Safety_Solutions.pdf

Transportation Information Center (TIC), Safety Evaluation for Roadways (SAFER) Manual
http://epdfiles.engr.wisc.edu/pdf_web_files/tic/other/SAFER_96.pdf

FHWA, Maintenance of Signs and Sign Supports
safety.fhwa.dot.gov/local_rural/training/fhwasa09025/fhwasa09025.pdf

FHWA, Proven Countermeasures
http://safety.fhwa.dot.gov/policy/memo071008/
http://safety.fhwa.dot.gov/provencountermeasures/

FHWA, Road Safety Audits
safety.fhwa.dot.gov/rsa/

FHWA, Road Safety Audit Guidelines
safety.fhwa.dot.gov/rsa/guidelines/documents/FHWA_SA_06_06.pdf

FHWA, Tribal Road Safety Audits: Case Studies
safety.fhwa.dot.gov/rsa/tribal_rsa_studies/tribal_rsa_studies.pdf

NCHRP 500 Series (Guidance for Implementation of the AASHTO Strategic Highway Safety Plan Transportation Research)
http://www.trb.org/main/blurbs/152868.aspx

Crash Modification Factors (CMF) Clearinghouse
www.cmfclearinghouse.org

Highway Safety Improvement Program (HSIP)
safety.fhwa.dot.gov/hsip

FHWA, Intersection Safety Toolkits
www.fhwa.dot.gov/resourcecenter/teams/safety/saf_3IST.pdf

Noteworthy Practices: Addressing Safety on Locally-Owned and - Maintained Roads – A Domestic Scan
safety.fhwa.dot.gov/local_rural/training/fhwasa10027/

Local and Rural Roads Safety Peer-to-Peer Program
safety.fhwa.dot.gov/local_rural/training/p2p/

Manual on Uniform Traffic Control Devices (MUTCD)
mutcd.fhwa.dot.gov

FHWA, Work Zone Law Enforcement Pocket Guide
safety.fhwa.dot.gov/wz/law_enforce/
AASHTO, Highway Safety Design and Operations Guide
safety.transportation.org/prgpub.aspx?pid=971
Tribal Highway Safety Improvement Program Model and Implementation Plan for Hazard Elimination Projects Guide
http://www.fhwa.dot.gov/tribal/topics/safety/saf_ack/
Building Tribal Traffic Safety Capacity
http://www.azdot.gov/TPD/atrc/Publications/project_reports/PDF/AZ592.pdf
Tribal Traffic Safety Funding Guide
http://www.azdot.gov/TPD/atrc/Publications/project_reports/PDF/AZ592s.pdf
Building Technical Capacity for Improved Tribal Consultation and Communication
http://www.tribalplanning.fhwa.dot.gov/caseStudies/arizona.pdf
Roadway Safety Noteworthy Practices Database
http://rspcb.safety.fhwa.dot.gov/noteworthy/
NHTSA Countermeasures that Work
http://www.nhtsa.gov/staticfiles/nti/pdf/811444.pdf
AASHTO Strategic Highway Safety Plan
http://safety.transportation.org/
Local Rural Road Owners Manuals: Intersection Safety
http://safety.fhwa.dot.gov/local_rural/training/fhwasa1108/
Roadway Departure Safety
http://safety.fhwa.dot.gov/local_rural/training/fhwasa1109/
Safety Information Analysis
http://safety.fhwa.dot.gov/local_rural/training/fhwasaxx1210/

## End Notes

[1] FHWA Highway Statistics (2009) http://www.fhwa.dot.gov/policyinformation/statistics/2009/hm20.cfm
[2] National Highway Traffic Safety Administration, National Center for Statistics and Analysis, Traffic Safety Facts, Rural/Urban Comparison, 2008 Data (PDF), DOT HS 811 164.
[3] Highway Statistics 2009 – Functional System Travel http://www.fhwa.dot.gov/policyinformation/statistics/2009/ vm2.cfm
[4] National Highway Traffic Safety Administration, Traffic Crashes Take Their Toll on America's Rural Roads (PDF), 2006.
[5] Highway Safety Improvement Program http://safety.fhwa.dot.gov/hsip/

[6] NCHRP Report 501 Integrated Safety Management Process is available online at http://onlinepubs.trb.org/onlinepubs/nchrp/nchrp_rpt_501.pdf

[7] http://www.deldot.gov/information/community_programs_and_services/DSHSP/index.shtml

[8] Strategic Highway Safety Plan (SHSP), FHWA Safety Program http://safety.fhwa.dot.gov/hsip/shsp

[9] http://safety.fhwa.dot.gov/local_rural/training/fhwasaxx1210/

[10] http://safety.fhwa.dot.gov/rsa/

[11] For more information on these countermeasures and strategies, consult the National Cooperative Highway Research Program (NCHRP) 500 Series Guidance Documents, available at www.safety.transportation.org, or the Federal Highway Administration's (FHWA's) Crash Modification Factor (CMF) Clearinghouse at www.cmfclearinghouse.org. Ideas for education and enforcement countermeasures can be found in NHTSA's Countermeasures that Work. Tools, training, guidance, and countermeasures for rural and local governments can be found on FHWA's website http://safety.fhwa.dot.gov/local_rural/training/. Several manuals for local rural road owners are available that address intersection safety, roadway departure safety, safety information analysis, speed management and non-motorized transportation

[12] See "A Primer on Safety Performance Measures for the Transportation Planning Process" at http://safety.fhwa.dot.gov/hsip/tsp/fhwahep09043/.

[13] AASHTO SHSP: http://safety.transportation.org/

[14] Countermeasures That Work: A Highway Safety Countermeasure Guide for State Highway Safety Offices (http://www.nhtsa.gov/staticfiles/nti/pdf/811444.pdf).

In: Local and Rural Road Safety
Editor: Mitchell Benson
ISBN: 978-1-63321-170-4


*Chapter 2*

# NOTEWORTHY PRACTICES: ADDRESSING SAFETY ON LOCALLY-OWNED AND MAINTAINED ROADS – A DOMESTIC SCAN*

***Rosemarie Anderson, P.E., Karen Yunk, P.E, Daniel Lovas, P.E. and Michelle Scism***

## ABSTRACT

Locally-owned road safety remains a challenge to many States. Several States have shown measured success in addressing local road safety. Seven States were identified to participate in the Local Road Safety Domestic Scan, allowing a team of transportation professionals from the Federal, State, and local levels to visit and document their practices. The Domestic Scan report identifies and documents practices in the planning, programming, and implementation of efforts to improve local road safety. Practices are presented in data collection and analysis; local project identification; local project administration; funding; training and technical assistance; outreach and partnerships between State Departments of Transportation (DOTs) and local agencies. The report provides States with valuable information to launch a local road safety program or implement documented practices to improve an established program.

* This is an edited, reformatted and augmented version of a document (Report No. FHWA-SA-10-027) prepared under contract for the Federal Highway Administration, Office of Safety, dated August 2010.

## EXECUTIVE SUMMARY

Locally-owned roadways in the United States are operated by more than 30,000 local jurisdictions covering approximately three million miles of roadway. The Fatality Analysis Reporting System (FARS) indicates that approximately 40 percent of all fatal crashes on the nation's highways occur on local roads. The amount of local roadway mileage and diversity of authority over local roadways create challenges for Federal, State, and local safety stakeholders to effectively direct funding and resources to mitigate safety issues on these roadways. Safety improvements at the local road level must be addressed methodically along with those efforts at the State level in order to significantly reduce the number of roadway related crashes on the nation's network. However, in many States local road safety still remains an afterthought due to lack of communication and/or resources.

The Federal Highway Administration conducted a Local Road Safety Domestic Scan of seven States to identify and highlight some of the most noteworthy programs, practices and procedures used by State and local agencies to address local roadway safety. The Local Road Safety Domestic Scan identified practices in funding, coordination, and technical assistance between State Departments of Transportation (DOTs) and local agencies with other States. States can use this information to launch a local road safety program; or identify practices that will improve an established program.

Six common themes emerged from the Local Road Safety Domestic Scan: data collection and analysis, local project identification, local project administration, funding, training and technical assistance, and outreach and partnerships.

Crash data remains the foundation of roadway safety management programs. Local road agencies need access to a comprehensive crash database to identify locations with safety issues to inform countermeasure selection. In an effort to build a comprehensive crash database that includes local road data, the Illinois DOT allocated Highway Safety Improvement Program (HSIP) funds to county governments to collect and geo-code their roadway crash data. The development of data analysis software with tools providing local agency access and training have proven effective in encouraging local road practitioners to take an active role in identifying safety issues and seeking resources to address them.

The scan States allocated all of their High Risk Rural Road Program (HRRRP) funds to the local road network; however, they all varied in how other Federal and State safety funds were used for local road projects.

Washington State DOT dedicated all their HSIP funds to the local road network in the State. Minnesota DOT distributes their HSIP funds based on the distribution of fatal and serious injury crashes regardless of road ownership. Scan States also provided some extra financial incentives to local agencies to solicit their participation in the safety programs.

Several of the scan States have streamlined the process by which local agencies obtain safety funds. For example, Minnesota DOT requires only one application for all sources of safety funds.

Additionally, Minnesota DOT uses visual reference tools to assist applicants in the selection of eligible projects for funding. Illinois DOT developed an online HSIP application tool to reduce preparation time and provide quick feedback to applicants regarding application status. Additionally, Illinois DOT provides workshops during their application process to train local agency personnel on safety priorities and techniques as well the application process. Many of the scan States have instituted other time-saving procedures including the use of force account, categorical exclusions and lump sum programming.

The main driving force behind methodologies used in the scan States to target local road projects was to identify projects that will most effectively address the State's safety goals. Often States require local road safety projects to address key area strategies or actions in the State DOTs Strategic Highway Safety Plan (SHSP) or County Road Plans. Washington DOT uses a corridor approach, where corridors are identified through their data-driven Corridor Safety Program. Both systemic and hot spot projects were encouraged by scan States.

Training, as well as partnership and outreach efforts targeted for the local practitioner are an integral part of the scan States' safety programs. The Local Technical Assistance Programs in all the scan States played a major role in offering safety training to local practitioners. Alabama DOT requires all county engineers to participate in data analysis and low cost safety improvement training to be eligible for HRRRP funds. In addition, the Illinois DOT provides road safety audit training for law enforcement officers.

Metropolitan Planning Organizations (MPOs) are intimately involved in the transportation safety programs in New Jersey, Michigan and Illinois. The MPOs foster partnerships and outreach with local agencies. This is accomplished by providing training and technical assistance programs to afford local agencies the knowledge and technical ability to develop successful local roadway safety programs.

Many State DOTs have dedicated personnel that provide oversight and assistance to local agencies, particularly with administration of Federal-aid projects. Visible local-aid divisions within the State DOT ensure relatively consistent coordination and collaboration between DOT personnel and local practitioners. Washington State DOT, Minnesota DOT and Alabama DOT each have a safety engineer dedicated to local agency technical assistance. In the other scan States, the safety office works closely with the local-aid division to promote and encourage local agency safety activity.

A local safety program can be used to advance State and regional priorities. A safety program that actively integrates local road safety improvement through funding of local safety projects, training for local practitioners, and outreach and coordination can assist States in achieving their safety goals.

# SECTION 1. INTRODUCTION

## Background

While the State highway and interstate roadway networks have historically accommodated the greatest traffic volumes and highest posted speeds, a review of the Fatality Analysis Reporting System (FARS) indicates that approximately 40 percent of all fatal crashes on the nation's highways occur on local roads. Local roads, as referenced in this document, are those owned and maintained by local jurisdictions (e.g., city, town, borough, county) and can be rural or urban. Local road agencies own and maintain approximately 75 percent of the nation's 8.4 million miles of roadway.[1] With continued urban sprawl and new mobility demands placed on many rural and suburban areas, safety challenges have grown on local roadways.

States are challenged to integrate road safety measures at the local level. To significantly reduce the number of roadway related crashes on the nation's network, safety improvements at the local road level have to be addressed methodically along with those efforts at the State level. However, in many States, local road safety still remains an afterthought due to lack of communication and/or resources. All levels of government are concerned that local road safety is not making the progress it should and Federal safety funds are not readily accessible to local agencies for safety improvements.

A number of issues inhibit the effective implementation of comprehensive and meaningful local road safety practices. Challenges to effective local roadway safety practices, include the following:

- The extent of the local roadway network.
- The variety of agencies involved in local roadway safety throughout each State.
- The lack of complete, accurate crash data and analysis tools.
- The diversity in organizational structure and capabilities of local jurisdictions.
- The limited training and expertise in roadway safety at the local level.
- The shortage of funding and sometimes relative low priority for a local safety program.
- Management structure that cannot adequately accommodate the delivery of a local road safety program.
- Federal requirements that may increase the cost and time to implement effective safety improvements.

States are required to address roadway safety on all public roads. In 2005, the Safe, Accountable, Flexible, Efficient Transportation Equity Act: A Legacy For Users (SAFETEA-LU) legislation created the Highway Safety Improvement Program (HSIP) as a core Federal-aid highway program. As States have focused on implementing the HSIP and achieving the crash reduction goals outlined by the Strategic Highway Safety Plan (SHSP), some States have developed new strategies for comprehensive and data-driven analysis; policies to facilitate local agency participation; procedures to allocate funding; and coordination efforts with local agencies to ensure effective implementation of local roadway safety programs.

## Scan Goals Objectives

Several States have shown success integrating safety at the local road level and streamlining local agencies' safety efforts. Seven such States, representing diverse geographic locations, were identified to participate in the Local Road Safety Domestic Scan, allowing a team of transportation professionals to visit and document their practices.

The goal of the domestic scan was to identify and document noteworthy practices in the planning, programming and implementation of efforts to improve local road safety. Noteworthy practices include coordination efforts between State and local agencies, such as technical assistance, and funding the implementation of safety projects and programs.

The objective was to share identified noteworthy practices in funding, coordination, and technical assistance with States and local agencies. By sharing information, States will gain insights on how to launch a local road safety program; or identify practices that will improve an established program.

## Scan Scope

The domestic scan involved a tour of seven States by a team of transportation professionals from the Federal, State, and local levels. The seven States visited were Alabama, Georgia, Illinois, Michigan, Minnesota, New Jersey, and Washington. Scan States were selected based on geographic diversity and innovative approaches to addressing local road safety. The following highlights some of the specific practices that led to the selection of each State for the scan:

- Alabama requires counties to participate in roadway safety training to be eligible for Federal funds.
- Georgia funds off-system (local) coordinator positions as well as funding for off-system projects.
- Illinois provides HSIP funds to local agencies to collect and geo-locate crash data and conducts safety workshops that highlight the application process for safety funds.
- Michigan's Local Safety Initiative provides technical assistance to local agencies.
- Minnesota develops county-level road safety plans to encourage low-cost countermeasures and creates funding targets for local agencies to use HSIP funding.
- New Jersey's local safety program is administered through regional planning agencies.
- Washington dedicates nearly all of its HSIP funding to local roadway safety.

At each of the seven States, round-table interviews were conducted with representatives from Federal, State and local transportation safety stakeholders. State DOTs and counties were the main participants in the domestic scan interviews, with representatives from Metropolitan Planning Organizations (MPOs), law enforcement, Local Technical Assistance Programs (LTAP), and Governor's Highway Safety Offices participating in some States. Participants from each State represented diversity in agency size, character of localities, and experience.

As outlined by the scope, this scan focused only on a small sub-set of the States conducting Local-aid programs for roadway safety. This report does not suggest that the programs and approaches used in the scan States comprise all of the necessary strategies for reduction of fatal and severe injury crashes on local roadways.

## Team Composition

The Scan Team was composed of representatives from the FHWA, American Association of State Highway and Transportation Officials – Standing Committee on Highway Traffic Safety (AASHTO-SCOHTS), the American Public Works Association (APWA), National Association of County Engineers (NACE), and Local Technical Assistance Programs (LTAP). The breadth and diversity of experience of the Scan Team representatives ensured a broad range of topics were covered and thoughtful discussions of noteworthy practices were completed during interviews with each State. Vanasse Hangen Brustlin, Inc. (VHB) served as the Scan Team facilitator and logistics coordinator. The full list of the Scan Team members and participants is included in the Appendix.

## Organization of Report

Six common themes emerged during the discussions and review of the domestic scan information provided by each of the States. These themes encompass a variety of similar practices that are used successfully in most or all of the scan States to improve local road safety. This report is organized by these themes:

- Data Collection and Analysis
- Local Project Identification
- Local Project Administration
- Funding
- Training and Technical Assistance
- Outreach and Partnerships

The remaining sections are dedicated to these common themes, citing noteworthy practices from the scan States and the benefits realized. The noteworthy practices are followed by notable local agency success stories, general observations and conclusions.

## SECTION 2. DATA COLLECTION AND ANALYSIS

Access to crash data and data analysis tools are the basis for an effective local road safety program. Crash data are used to identify locations with safety issues and assist in the prioritization of safety projects or programs competing for limited resources. However, many local agencies do not have access to crash data collected and compiled by their State or the technical training to analyze crash data and trends. As a result, local agencies with little or no access to crash data and analysis tools struggle to identify and address locations with data- driven safety needs.

To address this challenge, scan States utilize database, analysis, and mapping tools that can be easily shared with local agencies. They also provide technical assistance to aid local road practitioners in their safety analyses and countermeasure development. State assistance for local agencies in data collection and analysis can improve the ability of local agencies to develop projects and apply for funding.

### Data Collection

Local agencies need access to crash data to address safety on the roads they own and operate. However, obtaining comprehensive local road crash data has presented a challenge to some States. The Illinois Department of Transportation (IDOT) recognized that limited access to crash data is a barrier to identifying safety concerns that contribute to fatal and serious injury crashes on local roads. In 2005, IDOT embarked on an innovative program to improve their local road crash database.

Using HSIP funding, Illinois counties contracted for services to geo-locate five years worth of fatal and severe crash data on local roads in each county. Some counties pooled their funds and coordinated with a regional Metropolitan Planning Organization (MPO) to do the work. Other counties used temporary staff, part-time interns, or contractors to complete the work. Many local agencies recognized the significant benefits of the data and have chosen to continue collecting and geo-locating crash data using other funding sources. These crash databases have helped the counties meet the IDOT requirement for five years of data to apply for HSIP funds.

Similarly, the Michigan Department of Transportation (MDOT) funds two co-op students to geo- locate crash data in the State database. New Jersey DOT (NJDOT) also employs students to geo- locate crash data. The use of students is a cost effective means to improve crash data collection efforts.

## Data Analysis Tools

Crash data analysis can be time consuming and tedious when conducted manually, requiring a large portion of limited resources. Additionally, specialized training may be required to conduct data analyses. The identification of safety issues and selection of potential projects relies on crash data analysis. Employing user-friendly crash data analysis software tools to enhance the process has resulted in greater participation of local agencies in safety programs. Without data analysis tools, local agencies are challenged to review historical crash trends; calculate crash rates and compare them to statewide rates; and determine benefits of strategies that address safety issues on the roadways they own and operate.

All scan States have either developed State-specific software tools, or use ones available from other States These analysis tools are designed to help local agencies address roadway safety challenges by providing detailed crash information and identifying high risk locations. Scan States that provide local agencies access to data analysis software tools have shown increased interest in safety programs by these agencies. Highlights of specific data analysis tools are presented below.

***Alabama***

- Alabama Department of Transportation (ALDOT) utilizes the data analysis software package developed by the University of Alabama – Critical Analysis Reporting Environment (CARE).
- CARE is a free, user-friendly, statistical analysis software package designed for problem identification and countermeasure development.
- ALDOT analyzes crash data using CARE for the entire road network. A link-node internal system is utilized to review crash data to determine the average crash rates on roadways for fatalities and incapacitating injuries.
- Local agencies can access CARE once a confidentiality agreement is signed.
- CARE is web-enabled and is being integrated with GIS.
- CARE has the ability to analyze data and generate reports as well as crash diagrams.

***Georgia***

- Georgia Department of Transportation (GDOT) also uses the CARE statistical analysis software package to analyze crash data.
- Some of the features of the Georgia version of CARE include crash diagrams, data mining, and narrative data searching.
- GDOT manages the crash data input and the University of Alabama maintains the software.
- Access to the CARE software is controlled by GDOT, but made available to local and regional agencies through the University of Alabama's website.

***Michigan***

- MDOT funds and administers RoadSoft, a roadway management system with a safety component for storing and analyzing crash data.
- RoadSoft was developed by Michigan's Local Technical Assistance Program (LTAP) center.
- This software tool is available free of charge to all local agencies and provides not only a traffic crash database, but also data support for network screening, diagnostic tools, intersection analysis tools, and trend analysis.

- The output capabilities include graphical schematics with summary data, spreadsheets and tables that are used by all agencies to conduct a safety analysis.
- MDOT provides support and maintenance of the software, continually refining the data to ensure the most reliable and up-to-date information is available to agencies. MDOT has used safety funding to enhance the functionality of the safety module, including enhanced mapping, crash trend analysis, intersection collision diagrams, and new crash ranking options.

***Minnesota***

- The Minnesota Department of Transportation (Mn/DOT) Office of State Aid maintains the Minnesota Crash Mapping Tool (MnCMAT) system. MnCMAT provides a traffic crash database and crash analysis tools.
- All local agencies have access to MnCMAT. The output capabilities include graphical schematics with summary data, spreadsheets and tables that are used by all agencies to conduct safety analyses.
- The Minnesota Department of Public Safety provides data to Mn/DOT who maintains an internal system known as the Traffic Information System (TIS) with up-to-date crash data. The TIS data is used to populate the MnCMAT system.

***New Jersey***

- New Jersey Department of Transportation (NJDOT) has partnered with the Transportation Safety Resource Center (TSRC) housed at Rutgers University to develop Plan4Safety, a web-based software tool to analyze crash data in geospatial and tabular formats.
- All local agencies have access to Plan4Safety. The application allows users to quickly filter and analyze crash data and simultaneously merge specialized data sources (e.g., roadway data, EMS records) with those records for a more focused, in-depth analysis.
- Other Plan4Safety capabilities include identifying crash clusters, performing frequency analysis, generating road histograms of crash locations, calculating crash rates, diagnosing and providing national best practice countermeasures for low-cost improvements.
- The tool integrates statewide crash data and roadway characteristic data, calculates statistical analyses, incorporates network screening

layers and models, and includes a visual analytical tool with accurate GIS location data.

- The GIS feature provides an interactive, visual overview of how many and what types of crashes are occurring at a given location. Satellite map views clearly show geographic details of crash sites, such as neighborhood characteristics, buildings, and intersection configurations.
- Query results are presented in automatically generated charts and tables on the output screen.
- The software also includes a module that allows users to apply for safety grants through the Governor's Office of Highway Safety.

## Technical Assistance

Limited resources may deter several local road agencies from collecting crash data and conducting crash data analysis. States have developed software analysis tools that have proven more efficient in analyzing crash data. For various reasons, not all States allow local agency access to their software tools. Those that have access to the tools may require training to realize the full benefits. Scan States provide local agencies with comprehensive data collection and analysis support in multiple forms, including but not limited to: in-depth data analysis, daily fatality reports, crash maps for law enforcement and training support.

### *In-Depth Data Analysis*

Many State DOTs have elected to conduct in-depth crash data analyses on all public roads and share these with local agencies in lieu of providing crash data analysis tools.

The Illinois Department of Transportation expanded its crash database to include local crash data and analyzes crash trends on local roadways. IDOT uses an electronic database known as Safety Data Mart (SDM) to maintain traffic crash data within the agency. Generally, crash data reports are provided to the local agencies through a request to the IDOT District Office. The SDM also provides the data for the development and distribution of the five percent report location maps and the maps that show the location of fatal and serious injury crashes. These maps are distributed for use by the IDOT District Offices and local agencies.

Legal barriers in the State of Washington prevent the Washington State Department of Transportation (WSDOT) from making raw crash data available to local agencies. Because crash data is not directly shared with local agencies, WSDOT conducts comprehensive data analysis for local jurisdictions on a county-by-county basis to identify high crash locations and the most prevalent crash types. The results are shared with local agencies. These types of direct technical assistance identify locations with safety issues, provide safety data for review, and aid identification of potential safety projects with little or no resource expenditure by the local agencies.

The Michigan Traffic Crash Facts website provides a yearly compilation and analysis of Michigan crash data for government agencies, the media, and the public. The data is provided by the Michigan Department of State Police from the Michigan Traffic Crash Forms (UD-10). The Michigan Traffic Crash Facts website is created and maintained by the University of Michigan Transportation Institute (UMTRI) for the Michigan State Police Office of Highway Safety Planning.

### *Daily Fatality Reports*

Well-coordinated data analysis support between State DOTs, law enforcement and local road agencies provides a conduit for joint review and timely response to fatal crash causes. Both the Georgia DOT and Illinois DOT issue daily fatality reports to their district offices via email or web- access. In Illinois, the District Office sends a "scrubbed" version of this report to the County Engineers in which the fatal crash occurred. The District Offices work with local agencies and local law enforcement to review the need for any remedial actions.

### *Crash Maps for Law Enforcement*

ALDOT currently develops color coded crash maps, which can be broken out by jurisdiction. As shown in Figure 1, these maps provide a visual of high crash locations color coded by segments using a link-node system. Segments show high crash locations and specify areas of severe crashes. Currently only some local roads are included in the system. ALDOT is nearing completion of a project that will update their link-node system to have the capability to reference all county roads.

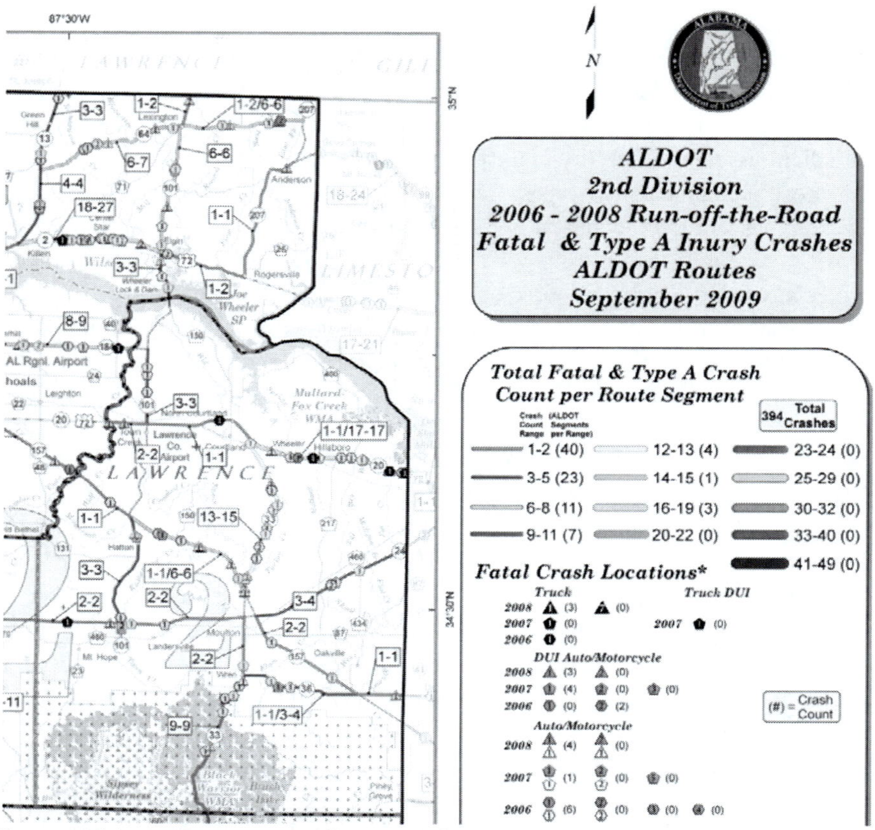


Source: Alabama Department of Transportation.

Figure 1. Example of Alabama Department of Transportation Crash Map.

These crash maps are used by several safety stakeholders including law enforcement. The police use the crash maps to develop plans to allocate manpower for traffic enforcement activity. Through the data gleaned from the maps the police can determine types of crashes, time of day and other crash attributes for selective enforcement.

### *Regional Crash Profiles*

Many State DOTs often work with regional agencies to address data analysis challenges for local practitioners. In Georgia and New Jersey, the regional MPOs play an important role in bridging the gap between the State DOT and local transportation agencies. For instance, the Atlanta Regional

Commission (the MPO for the Atlanta Metropolitan Area), develops crash profiles for member counties every two years. These crash profiles identify locations with high crash severity by location or victim type, including corridor, intersection, pedestrian, and bicycle locations.

In New Jersey, the North Jersey Transportation Authority (NJTPA) conducts data analyses that identify safety priorities for each county within its region. County transportation agencies review the maps to identify locations with safety issues that may be candidates for safety projects.

***Training Support***

In developing analytical software tools for local road agency personnel to utilize in their safety projects and programs, scan States have also recognized that training is a necessary component. To assist local agencies in understanding how RoadSoft can be used in their safety programs, MDOT sponsors multiple training sessions each year, geared toward basic, intermediate and advanced user levels. These training sessions are usually conducted by the Michigan LTAP center. The University of Alabama works with ALDOT and GDOT to provide training courses to local agencies for use of the CARE software. In New Jersey, the TSRC provides training to agencies on the use of the Plan4Safety software and its capabilities. These training courses help local agencies to fully utilize the software tools for data analysis independent of the State DOT.

## Summary

Crash data is the foundation for an effective transportation safety program. Crash data is used to identify locations with safety issues and assist in the prioritization of safety projects or programs competing for limited resources. Access to a comprehensive crash database and data analysis tools are major steps in advancing with a local roadway safety program. Many State DOTs have recognized the importance of placing crash data analysis and tools in the hands of local agencies. As a result:

- IDOT uses HSIP funds to collect and geo-locate crash data on local roads. This helps local agencies support their application for safety funds.
- MDOT capitalizes on relationships with law enforcement for timely submission of crash data.

- ALDOT, GDOT, MDOT, Mn/DOT, and NJDOT utilize database, analysis, and mapping tools that are shared with local agencies. These tools are the primary means by which local agencies identify locations with critical safety issues. These State DOTs also provide local road agencies with training on the use of data analysis software tools.
- IDOT and WSDOT provide in-depth data analyses for local jurisdictions. This assistance has led to the identification of safety issues, development of safety countermeasures, and applications for safety funding to complete improvements on local roadways in the States.
- IDOT, ALDOT and GDOT provide daily fatal crash reports to local stakeholders to enable quick response to safety issues.
- MPOs in Georgia and New Jersey develop regional crash data maps or profiles that local agencies use to assess locations with safety concerns.

State assistance for local agencies in data collection and analysis improves the ability of local agencies to develop projects and apply for funding in the scan States. When State DOTs provide the means to better collect, analyze, and understand crash data, they expand and improve the ability for local agencies to make valuable decisions regarding local roadway safety. These activities provide the foundation for local agencies to identify projects that will provide the greatest benefits to transportation safety in their region.

## SECTION 3. LOCAL PROJECT IDENTIFICATION

A consistent, organized and well-communicated project identification process is needed for a successful local road safety program. The scan States have instituted policies and procedures to direct local agencies to identify projects that address their roadway safety needs and support statewide crash reduction goals. The local roadway safety projects may take many forms, range widely in cost, and include both hotspot improvements and broader systemic safety measures. This section describes some of the approaches to project identification used in the scan States.

## Strategic Highway Safety Plans

State DOTs have developed Strategic Highway Safety Plans (SHSP), which identify the State's safety goals, priorities and strategies. Local agencies are often directed to identify projects based on the priorities established in the SHSP. Additionally, several scan States encourage low-cost treatments that can be applied systemically to address overrepresented crash types or patterns.

MDOT's HSIP process prioritizes broad systemic treatments that align with the priorities and recommendations included in the Michigan SHSP. MDOT compiles and distributes a resource list that indicates high-priority crash types, based on emphasis areas of the SHSP, and potential corridor-wide countermeasures, which are considered effective and likely to receive approval for safety funding. MDOT encourages local agencies to address locations in the State's Transparency Report and maintain consistent countermeasures throughout the roadway system. The final decision on project identification is left to the local agencies. Safety initiatives have included guardrail upgrades, signal modernizations, and corridor signal timing upgrades, which all tie into the lane departure and intersection focus areas of the Michigan SHSP.

Washington State Department of Transportation (WSDOT) reviews and updates their SHSP priorities every two years identifying priority crash types statewide. Priority crash types are identified based on fatal and serious injury crashes statewide. Roadway departure crashes have been elevated to the highest priority for remedial actions based on the most recent statewide crash data review. Once SHSP priorities are set, WSDOT conducts statewide analyses of local roadways on a county-by-county basis to identify the top fatal and severe injury crash locations. WSDOT determines the top 25 percent of counties (equivalent to 10 counties) that qualify as "above average" for a priority collision type.

## County Road Safety Plans

Many local agencies would benefit from their State's assistance and guidance to assess local roadway safety and identify projects related to the SHSP goals. Minnesota developed a County Road Safety Plan (RSP) Program to help its counties identify potential safety projects. The program uses funding from traffic violations and takes a 4E (engineering, education, enforcement, emergency service) approach.

The RSP process enables State safety specialists to work alongside county staff and other safety stakeholders to identify a prioritized list of projects. Mn/DOT, via a consultant, provides technical assistance to counties to identify local safety emphasis areas, priority strategies, and a prioritized list of safety projects. The primary objective of the RSPs is to identify a specific set of low-cost, systemic safety projects that are linked directly to the causation factors associated with the most severe crashes on the county's roadway system. Minnesota intends to implement RSPs in all counties within three years.

The RSP process increases safety awareness among local staff and stakeholders. Local agency personnel are more familiar with identifying safety issues on their roadways and suitable countermeasures to address them.

## Corridor Safety Program

Local agencies are often intimidated by the volumes of crash data that must be reviewed to properly identify safety issues on their roadways. Some local agencies are unsure what process to use to translate crash data for an entire jurisdiction into the development of individual safety projects. In a combined effort with the Washington Traffic Safety Commission (WTSC), WSDOT has developed a Corridor Safety Program that identifies high crash corridors on the local road network. This initiative relieves local agencies of the data analysis requirements and helps local agencies address safety issues on their roadways.

The high-incidence corridors are selected based on crash analyses showing a higher than average rate of crashes. The length of corridors for this program ranges significantly, from 3.5 to 60 miles in length. A task force is formed and site visits are conducted to identify specific safety-related issues on selected corridors. Suitable low-cost safety countermeasures are identified and implemented to improve safety on the corridor. The typical duration of review and improvement implementation on a selected corridor ranges between 18 and 24 months. WSDOT has used marketing efforts, such as branding, billboards, and websites, to publicize the projects and their safety benefits to roadway users. The Corridor Safety Program has been successful at increasing road safety in addition to building community relationships and safety awareness.

## Regional Programs

The New Jersey Department of Transportation (NJDOT) established the Local Safety Program to provide funding for local road safety projects. Technical assistance and outreach were needed to promote this new program and educate local agencies on safety management practices. However, NJDOT recognized that they did not have the resources to support the program in this manner. Therefore, NJDOT partnered with the MPOs to identify and select local safety projects.

NJDOT regularly conducts an analysis of the crash data using a web-based crash data application, Plan4Safety, to determine areas to concentrate resources. These locations are separated into lists according to roadway characteristics and shared with the respective MPO. Each MPO is allocated a portion of HSIP funds for local road safety projects. The three New Jersey MPOs then develop programs that best meet their region's needs. North Jersey Transportation Planning Authority (NJTPA) periodically conducts a study to identify regional safety priorities, identifying potential locations for improvement based on criteria such as degree of crash exposure and geographic distribution of crashes, with a focus on county and local roads. The South Jersey Transportation Planning Organization (SJTPO) sponsors road safety audits at locations identified through a data-driven process. The Delaware Valley Regional Planning Commission (DVRPC) works with their counties to identify projects from various planning studies to advance to implementation through the Transportation Improvement Program.

MPOs can provide valuable coordination and assistance to local agencies to address regional safety priorities. The New Jersey example capitalizes on existing relationships and organizational structure, while promoting transportation safety planning that develops programs and projects to meet the goals and objectives of the region.

## Visual Reference Tools

Rather than rely on local agencies to search through copious amounts of research and other sources of information on countermeasure effectiveness, Mn/DOT has developed an easily- accessible summary of preferred roadway safety treatments.

Mn/DOT created a visual reference tool, the HSIP Impact Pyramid, shown in Figure 2. The HSIP Impact Pyramid succinctly shows the relative benefits of various roadway safety measures by grouping individual countermeasures in a hierarchy of four "impact" tiers. The pyramid shows the most beneficial strategies on the largest tier (the pyramid base/foundation) and narrows to the least beneficial items on the smallest tier (the pinnacle). The HSIP Impact Pyramid reflects Mn/DOT's preference for systemic HSIP improvements that will result in the greatest impacts to local roadway safety, while acknowledging that reactive site-specific measures must also be considered. This tool has helped local agencies understand which improvements are effective, select eligible projects, and reduce crash potential on local roadways.

In addition, Mn/DOT uses what they refer to as the *Proactive Spectrum*, shown in Figure 3, in the project selection process for HSIP funding. A minimum of 70 percent of the projects awarded to each Area Transportation Partnership (ATP) are proactive; while a maximum of 30 percent of projects awarded to each ATP are reactive projects and must have a benefit/ cost (B/C) ratio greater than 1.0 to be considered for funding. The Proactive Spectrum is not inclusive of all potential safety strategies. Additional strategies may be appropriate for some roadways. Minnesota created the Proactive Spectrum to help decision-makers achieve the maximum safety benefit as intended by HSIP.

Source: Minnesota Department of Transportation State Aid Division.

Figure 2. Minnesota HSIP Impact Pyramid.

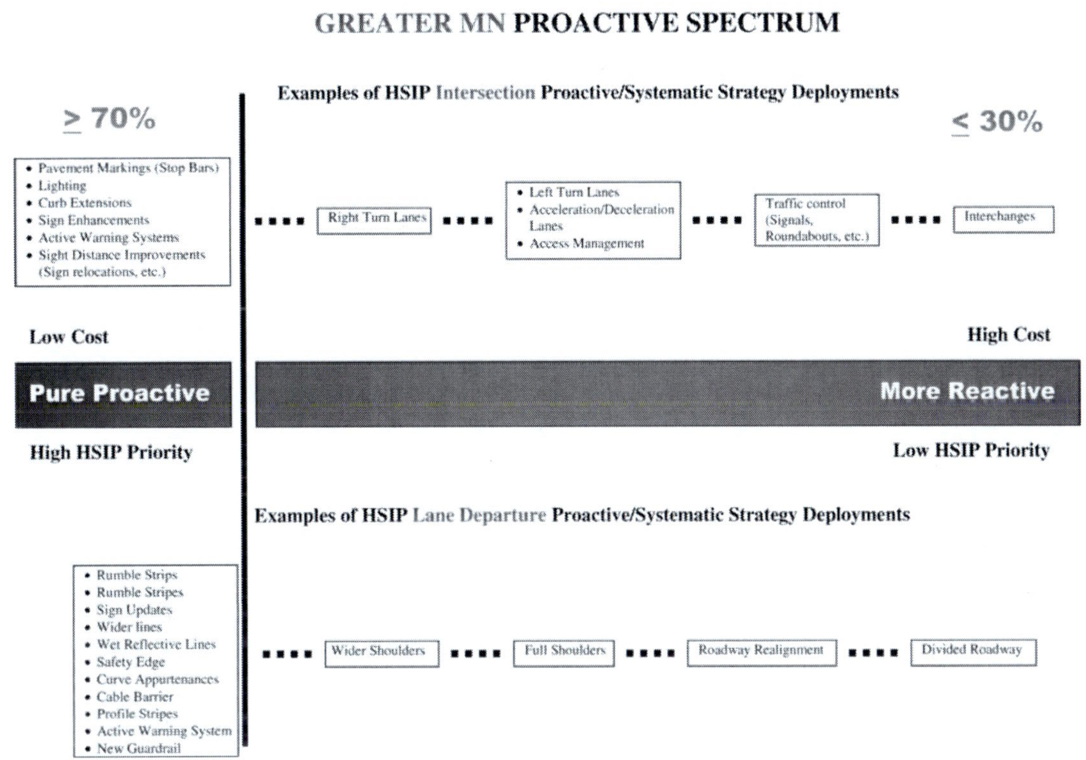


Note: The Proactive Spectrum is not all inclusive of all safety strategies. Additional strategies may be appropriate for some roadways.
Source: Minnesota Department of Transportation State Aid Division.

Figure 3. Minnesota HSIP Impact Pyramid.

## Summary

Policies and procedures developed by the scan States have helped local transportation agencies identify systemic and site-specific improvements to improve transportation safety. Scan States have identified priorities and developed mechanisms to steer local road agency project selection processes.

In summary, practices for identifying projects on local roadways in the scan States include:

- MDOT and WSDOT focus on priorities established in the State SHSP.
- Mn/DOT develops county road safety plans that identify low-cost, systemic safety projects that are linked directly to the causation factors associated with the most severe crashes on the county's highways.
- In NJ, MPOs provide valuable coordination and assistance to local agencies in data analysis and project identification to address regional safety priorities.
- The WSDOT Corridor Safety Program identifies low-cost countermeasures to address roadway segments with safety issues. Projects are locally managed with WSDOT and WTSC assistance.
- Mn/DOT uses visual reference tools to show the impact of countermeasures and give local practitioners guidance to select both proactive and reactive projects.
- States are looking beyond individual crash sites to implement systemic improvements on corridors and roadway networks that are susceptible to overrepresented crash types.

The project identification practices discussed in this section provide resources for local transportation agencies to effectively pursue transportation safety funding and allow the State DOTs to remain involved in the review and analysis process. Productive collaboration between transportation agencies at all levels lays the groundwork for successful project identification. States must remain involved and guide local agencies through the project identification process to realize safety goals.

# SECTION 4. LOCAL PROJECT ADMINISTRATION

Local roadway agencies across the nation have major personnel diversity in terms of number, expertise and duties. Large agencies have dedicated staff that serves as project administration liaisons with State DOT personnel. As a result, these individuals are well versed in the requirements for successfully implementing transportation projects utilizing State and Federal sources of funding. Dedicating administrative resources and committing technical staff to acquire funding and implement projects can be difficult for local agencies to manage.

To assist local agencies in completing safety projects, many State DOTs have allocated staff resources to coordinate with local agency staff to administer local safety projects. States have created separate departments to administer funding and technical assistance to local agencies. The departments create guidelines and procedures to guide local safety programs and encourage participation in the process. Many agencies have also instituted time-saving procedures to streamline the project development process. This section describes some of the practices related to the administration of local safety programs in the scan States.

## Local-Aid Divisions

Local transportation agencies vary widely in size, funding, technical capabilities, and administrative resources. Application for transportation funds in many States often involves significant effort, which can be a burden for many local agencies. As a result, local agencies may be reluctant to apply for available funds. State DOTs have created local roadway divisions to develop application guidelines, review project applications, provide technical assistance to applicants, and provide overall administration of local roadway improvement programs.

The DOTs in all scan States have a Local-aid division. While the structure and operation of the Local-aid divisions vary among scan States, the Local-aid division typically consists of a central office with support from district or regional offices. The central office is generally responsible for developing program applications and guidelines, as well as project selection, whereas the regional or district offices are involved in the day-to-day coordination with local agencies. Specific safety-related technical assistance is provided by State liaisons or MPOs. The State liaisons can be either dedicated staff of the Local-aid division or representatives of the DOT's safety office. The Local-aid division may also manage projects on behalf of local agencies. The greatest benefit of the Local-aid divisions is that they establish a formal mechanism for local agencies to obtain State and Federal funding.

The strength of these divisions, the breadth and depth of technical assistance, and the resulting relationship-building between the State DOT and the local agencies are significant factors in the success of the local road safety programs in the scan States.

## Application Administration

States have developed application procedures that provide clear guidance to local agencies seeking funds for safety projects. The scan States have developed preferred procedures for soliciting and reviewing applications seeking HSIP, High Risk Rural Roads Program (HRRRP), and State safety funding. State guidance includes project selection criteria practices and monitoring that encourage local agencies to complete the projects in a timely manner. The following describes practices in project application guidelines and program procedures that have assisted local agencies in obtaining and using funds for safety projects.

### *Application Committees*

ALDOT established a HRRRP Committee to seek local stakeholder involvement and input during the project selection and prioritization process for the High Risk Rural Road Program (HRRRP). This committee includes representatives from ALDOT, counties, and the FHWA Alabama Division. The primary purpose of the committee is to establish a competitive methodology for project eligibility and selection as well as guidelines and procedures to obtain HRRRP funds. Each year the eligibility requirements established by the committee are reevaluated to ensure that competitive funding supports projects designed to reduce crashes and maximize benefits on rural roadways.

ALDOT dedicates 100 percent of its HRRRP funds to the local road system. County input in the application process and guidelines ensures county engineers buy-in. This has enhanced participation in the HRRRP.

### *Application Tools*

The Illinois Department of Transportation has developed an online HSIP application tool, including a built-in benefit-to-cost (B/C) analysis function. The IDOT application requires a variety of information and data requirements for local agencies to complete, including photos, crash data analysis, and B/C analysis through an electronic online form. Once the application is completed, it is instantly available within the IDOT system for review. An IDOT review team initially reviews B/C analysis information for completeness and relevance before reviewing the entire application. Application issues can be identified and resolved quickly through the online process.

### *Staggered and Flexible Solicitation*

Local agencies unfamiliar with the Federal-aid program requirements may not understand which funding mechanisms best fit their local roadway safety projects. In addition, where multiple sources of funding exist, submitting an application for each can be time consuming. The time consuming process can act as a deterrent to submitting an application, especially when local agencies are competing for small amounts of funding.

MDOT's standard call for HSIP projects is completed on an annual basis. The annual requests for HSIP and HRRRP projects are staggered by several months to allow local agencies to respond to the more restrictive program (HRRRP) first. All local safety projects not receiving HRRRP funding are automatically rolled into the HSIP selection process, giving those projects a second chance at funding. This policy elevates the potential for each local project to receive funding.

Beginning in fiscal year (FY) 2010, MDOT now has the HSIP projects on a two year cycle (i.e., MDOT solicited FY 2011 projects in FY 2009) to give local agencies needed lead time to complete preliminary steps, such as topographical surveys, right-of-way (ROW) identification, conceptual design, and budgeting. The following summarizes the typical two-year program schedule:

- Application window: November 2008 – March 2009.
- Scoring: March 2009 – July 2009.
- Selection: July 2009 – August 2009.
- Authorization for Implementation: October 1, 2010.

The two-year cycle gives adequate time for local agencies to prepare their applications. The cycle also provides a consistent and reliable schedule for local agencies.

### *Universal Applications*

Limited time and resources limit the number of applications local agencies submit for safety project funding. Mn/DOT permits local agencies to submit one application for consideration under all Federal and State safety funding sources. Mn/DOT annually distributes funding solicitation notices to local agencies for eligible local roadway safety projects. Once the final list of projects is approved, the projects are reviewed for eligibility with appropriate funding sources. The single application reduces complexity for local applicants and minimizes the need to expend additional resources on multiple applications.

### *Joint Applications*

Historically, low levels of cooperation between local agencies leads some safety projects to terminate at jurisdictional boundaries or go undeveloped altogether. Mn/DOT encourages local agencies to combine efforts and submit joint applications through the Mn/DOT districts when possible. The county engineers agree to manage different phases of the project delivery process for these joint applications. For example, one county engineer might lead early efforts on coordinating and submitting an application for funding, while another county engineer would handle the later tasks of construction inspection, documentation, and contractor payments after a project has been funded and let for construction.

This process helps Mn/DOT allocate funding to multiple agencies and encourage partnership amongst neighboring communities. Mn/DOT's joint applications encourage consistent and region-wide responses to common safety issues, particularly when systemic safety improvements are implemented across jurisdictional boundaries.

### *Prioritization*

States with clear guidelines to prioritize projects generally receive well-prepared and targeted project applications. MDOT specifies the following elements will be used to evaluate and prioritize local safety projects:

- Scope of work.
- Time of Return (TOR).[2]

- Ratio of estimated cost-per-annual benefit.
- Ties to the SHSP initiatives.

These criteria ensure that local roadway projects with the greatest crash-reduction benefits receive the highest priority. Local agencies that consider these criteria in their applications stand the greatest chance for selection.

Projects that address locations in Michigan's five percent report are also evaluated using the criteria. MDOT finds that the the Time of Return analysis results are very good for safety project applications related to locations in the five percent report, which generally allows these projects to rise to the top in the overall scoring.

***Project Award Requirements***

While improved application procedures can increase local agency participation, States should ensure safety funds are spent efficiently. Workable safety programs require States to develop requirements that achieve the highest possible rate of project completion.

The IDOT process takes approximately two years from the solicitation to complete design and construction of individual safety projects. IDOT generally attempts to select HSIP and HRRRP projects by the beginning of the State fiscal year and strives to complete local roadway safety projects in the same fiscal year that the project funding is awarded. This requirement keeps the local agencies focused on applying for construction-ready projects and other projects that have received public support from local officials or stakeholders.

Like IDOT's policy, all MDOT local roadway projects must be completed in the same fiscal year that the project funding is awarded. MDOT gives agencies a notice to prepare for the funding in advance of the award. This allows ample time to prepare for the project so that the funding can be obligated as soon as possible after the award. MDOT also establishes a cap for Federal funding allocated to each project, which ensures funding can be awarded to a large and diverse pool of local recipients and limits the potential for any one recipient to receive, but fail to use, a significant portion of the funding.

## Streamlining the Federal-Aid Process

States are adopting processes and procedures to streamline their safety programs and expedite the process to encourage local road agencies to prepare and submit applications for safety funds. The following describes practices used by the scan States to streamline the safety program process.

### *Force Accounts*

The Federally required competitive bid process for construction projects is often burdensome and time consuming for local agencies, requiring up to two years before some projects are constructed. This represents a challenge, particularly for the small amounts of funding often associated with low-cost safety improvements.

The Alabama Department of Transportation permits local agencies to use force account[3] to enable delivery of small projects in order to save time on the overall process. MDOT also permits local projects to be constructed using force accounts. Federal agencies must document a finding of cost effectiveness and receive FHWA Division Office approval to use force account procedures. States have found that force account enables expedited delivery of small projects, such as pavement markings, at considerably lower cost than if implemented via the competitive bid process.

### *Abbreviated Plan Review Process*

The number of steps, level of information, and review time required to complete the Federal- aid application process can be discouraging. Local agencies would prefer to complete the process as quickly as possible to implement much needed improvements, particularly low-cost measures that can reduce crashes immediately.

The process for handling Hazard Elimination Safety Type (HES-TYPE) projects within the ALDOT Central Office has been streamlined. First, HES-TYPE projects must meet minimum eligibility requirements. Then, only project work directly related to the safety problem can be included in the application. Proposed projects normally should not require any right-of- way acquisition or major utility relocations. The intent of the streamlined review process is to produce fast track projects that will take the least number of steps in the plan preparation stage.

HES-TYPE projects are developed using a minimum number of plan sheets. The plan assembly for HES-TYPE projects should be held to a minimum, typically with 10 sheets or less consisting of a Title Sheet, Typical

Section Sheet, Project Note Sheet, Summary of Quantities, Plan-Profile Sheet, Traffic Control Plan (TCP), and Cross Sections. The designer can utilize special and standard drawings to complete the design and plan preparation effort. This reduces the plan review process by allowing for minimal reviews and early authorization.

### *Project Programming*

In order to receive HSIP funding for local roadway safety projects, the projects and their funding sources must be included in the Statewide Transportation Improvement Program (STIP). The STIP includes all capital and non-capital projects funded under Title 23, United States Code - Highways, covers a period of not less than four years, and is updated at least every four years. Highway safety improvement programs are often developed on an annual basis; therefore, it is sometimes difficult to know which projects will be advancing in four years. Fortunately, States can group projects by function, work type and/or geographic area in one line item if a project is not of regional significance, or of appropriate scale to warrant individual identification. For example, several low-cost safety enhancements could be grouped together and listed as various safety improvements for one county.

In order to provide sufficient flexibility with programming safety projects on an annual basis, Illinois uses a combination of line-item and project-by-project listing in the STIP. For State and local safety projects not within the borders of an MPO, IDOT uses a STIP line item and amends the list of projects based on their annual status. For safety projects within an MPO, the MPO is responsible for listing each project in their respective TIP. Use of STIP line items (i.e., lump sum programming) allows for greater flexibility in programming projects on an annual basis and addressing data-driven needs to move projects forward.

### *Categorical Exclusions*

Highway safety projects receiving Federal funding are required to complete environmental impact reviews. Local agencies planning to construct low-cost or systemic improvements, with little or no potential for environmental impacts, frequently cite these requirements as a deterrent to applying for funding.

Federal legislation for implementing the National Environmental Policy Act (NEPA) defines a number of Categorical Exclusions (CE) exempting certain types of safety projects from extensive Federal environmental impact review (i.e., preparation of Environmental Impact Statements). Projects qualify as CEs if they meet various requirements demonstrating a likelihood of minimal infrastructure, environmental, or historical resources impacts. For instance, the legislation exempts the construction of pedestrian/bicycle facilities, as well as the installation of signs, pavement markings and traffic signals (where no substantial land acquisition or traffic disruption will occur) , among others,[4] from further NEPA review.

New Jersey's Local Safety Program identifies several project selection criteria. One of these criteria is the requirement that minimal or no environmental issues exist, which allows the project to obtain a CE classification. A CE classification requires significantly less time and resources than environmental impact statements or environmental assessments, thereby resulting in expedited project delivery.

Strategies designed to simplify or shorten the Federal-aid process have allowed local agencies to increase participation and complete more local safety projects.

## Summary

Many State DOTs have dedicated personnel that provide oversight and assistance to local agencies in their respective States particularly with administration of Federal-aid Projects. Visible Local-aid divisions within the State DOT ensure relatively consistent coordination and collaboration between DOT personnel and local practitioners. The primary role of the Local- aid divisions is to provide guidance to local agencies to navigate the application process and implement projects. Specific technical assistance is often provided by subject matter experts. For example, WSDOT, Mn/DOT and ALDOT each have a Safety Engineer dedicated to local agency technical assistance. In the other scan States, the Safety office works closely with the Local-aid division on local agency safety activities.

While the structure of the Local-aid divisions varies across the States, the benefits to local agencies are mirrored in all the scan States, as follows:

- Guidance (often hands-on) to navigate the State's application process and implement projects.
- Technical assistance to local agencies in identifying and prioritizing projects with the greatest safety benefits.
- Strong partnerships between agencies are established, which enhance their ability to achieve statewide crash reduction goals.

- MDOT and ALDOT use force account procedures that allow local agencies to use internal resources to construct projects, which saves time and resources on the completion schedule.
- IDOT programs safety projects by consolidated line items in the STIP to allow flexibility in moving projects forward.
- Categorical exclusions may be used to exempt local safety projects with minimal environmental impacts from Federal environmental review, which speeds the approval and completion schedule.

Many of the scan States have instituted innovative practices to administer local safety programs and streamline the Federal-aid process to encourage greater local participation. Some examples include:

- ALDOT formed a committee with local agency representatives to develop methodology, and application guidelines for safety programs.
- IDOT developed an online HSIP application tool to reduce production times and provide quick feedback to applicants if their application is incomplete.
- MDOT instituted a two year cycle for application to ensure completion of projects in the same year funds are awarded.
- Mn/DOT requires one application for all safety funding sources.

The scan States' Local-aid administration structures and practices may serve as models for other States to develop effective management of local safety programs. Dedicated State DOT local-aid divisions and application guidelines facilitate greater local agency participation and more effective projects. These divisions understand the various funding mechanisms for local safety projects and provide instrumental support for local agencies to acquire and obligate funding.

## SECTION 5. FUNDING

Most local jurisdictions do not have dedicated transportation safety funds. As a result, safety projects have to compete with other local transportation needs for funds. Often the competition for funds is not limited to transportation needs but other local requirements like schools and utilities. Without safety funding from outside sources, local agencies may be unable to complete essential safety projects. In some cases, Federal and State programs set aside money to distribute to local roads.

SAFETEA-LU established the HSIP, which provided over $1 billion annually through FY 2009 to fund highway safety projects on all public roads. The HRRRP, which is a set aside component of the HSIP funds, supports safety projects on lower functional-class rural roads. The scan States have developed HSIP funding mechanisms to increase local agency participation in the Federal- aid process and to distribute funding to eligible safety projects. The following describes some of the funding practices used by scan States to encourage local agency participation in their safety programs.

## Funding Distribution

The approach to funds distribution to local agencies varies amongst the scan States. Some States have chosen to allocate a significant portion of HSIP and HRRRP to local agencies. In most instances, this funding is distributed through State DOT districts or regional MPOs.

### *Federal Safety Funding*

Local agencies require Federal or State funding support to implement local roadway safety projects. State DOTs have to balance the safety needs of the State-owned and operated system with that of the locally-owned and operated network within available funds. Often States use the allocation of HSIP funds as leverage to influence local agencies' participation in State safety programs that will address statewide safety goals.

ALDOT dedicates 100 percent of its HRRRP funds to county and municipal rural road safety projects. ALDOT also flexes approximately 10 percent of their HSIP funds to law enforcement for statewide traffic safety enforcement. Through crash data analysis, ALDOT advises the law enforcement agencies on safety areas of emphasis, and locations for targeted enforcement. ALDOT's distribution policy supports a balanced multi-disciplinary approach to reducing roadway crashes.

In New Jersey, the Local Federal Safety Program provides a minimum of $1 million to each of the three MPOs, dedicated to local road safety projects. These funds are earmarked for construction only. This funding for local agencies is managed by the MPOs through the Transportation Improvement Program (TIP) process.

The IDOT Office of Programming controls the distribution of HSIP funds for local agencies and then informs the Local Roads and Streets Division of the amounts to distribute to each agency. The Local Roads and Streets Division is allocated 100 percent of the State's HRRRP funds and 20 percent of the State's overall HSIP funding to distribute to local roadway safety projects. MDOT allocates all HRRRP funds and 30 percent of the HSIP funds to local roadway safety projects. These funding levels represent a strong commitment from IDOT and MDOT to assist local agencies in addressing safety issues on local roadways.

WSDOT channels nearly 100 percent of its HSIP funds to local transportation safety projects through the WSDOT Highways & Local Programs Division.

### *Data-Driven Allocation Policies*

Not all States devote a percentage or all of their HSIP funds to local safety projects. Mn/DOT uses a data-driven methodology to allocate funding statewide based on crash rates. Mn/DOT distributes their HSIP funds to the districts based on the percentage of statewide fatal and serious injury crashes occurring in each district. Further allocation of HSIP funds at the district level for State and local roadways is based on the percentage of fatal and serious injury crashes occurring on each system. This policy provides the advantage that the distribution of HSIP funds is directly related to the distribution of crashes on the State and local roadway system.

### *Other Funding Sources*

HSIP funds are limited in scope and availability. States can use other sources of funding to supplement local needs not covered by HSIP funding.

The New Jersey MPOs allocate some Federal-aid planning funds each year to develop and prioritize safety projects. These projects include conducting Road Safety Audits (RSA), safety planning studies and coordinating Regional Safety Task Force activities. HSIP and other local safety projects are developed directly from RSAs conducted with this funding.

Mn/DOT has developed separate State funding mechanisms to support local roadway safety. The Rural Road Safety Account uses general State funds to fund projects. In 2008-2009, approximately $10 million was allocated to local agencies for rural highway safety projects that focused on reducing fatal and serious injury crashes. These Rural Road Safety Account funds may only be used for construction; the counties must fund the design services, right-of-way acquisition, and may contribute to construction costs as well. This account is considered an essential supplemental source of funding to local agencies.

Minnesota has also created a Central Safety Fund, comprised of Federal Section 164 repeat traffic offender sanction monies. In 2006, DUI and seatbelt violation monies were also added to the Central Safety Fund. These funds are typically directed towards additional law enforcement on local roads. Historically, a substantial proportion of the Central Safety Fund has been used to support local roadway safety projects throughout the State.

## Financial Incentives

The Federal requirement for matching funds for some types of safety improvements can directly impact the types of projects selected for implementation. Local agencies without the available matching funds are limited to those safety countermeasures that are 100 percent eligible for Federal funds. This can reduce local agency participation in the State's safety programs. To counteract this deterrent, States have developed financial incentives to encourage local agency participation in highway safety programs.

### *State Match Contribution*

Federal-aid funding programs typically cover 80 to 90 percent of project costs, with the remaining 10 to 20 percent to be provided by another source of funds. However, for safety projects that are not 100 percent eligible for Federal-aid, States and local agencies are required to provide 10 percent of the project cost.

The Georgia Department of Transportation learned from previous experience with their Off- System Safety Program, that many local agencies were unable to provide matching funds for safety projects. Local communities could not obtain sufficient project funding to make potential safety projects viable. For the HRRRP, GDOT supplements the Federal funds with State funds to provide local agencies with 100 percent of required cost for eligible safety projects. This incentive has increased local agency participation in the program and helps to ensure that safety projects are initiated and completed.

The Washington Department of Transportation has also implemented a similar monetary incentive to encourage participation from local agencies to identify and implement safety projects. WSDOT has increased local agency participation in the project application process by providing the 10 percent matching funds for HSIP-eligible local roadway safety projects.

***Preliminary Engineering***

Limiting Federal HSIP funding to construction costs for safety projects leaves the responsibility for both the application and design costs to the local agencies. Some State DOTs and regional planning agencies support non-construction costs to reduce the financial burden on local agencies and meet agency goals.

Projects eligible for NJDOT's Local Safety Program can be awarded funding for construction only. Many local agencies need assistance for project design and/or application preparation. The South Jersey Transportation Planning Organization (SJTPO) has developed a process to encourage local agencies to implement road safety audit (RSA) findings. The MPO funds contractor support to assist local agencies with project development efforts, including preliminary engineering (PE) and preparing the Federal-aid application package. This practice has advanced SJTPO's RSA program and encouraged local agency participation in the State's Local Safety Program.

Similarly, WSDOT provides 100 percent of the funding required for the PE and final design of local roadway safety projects. WSDOT contributes the costs to acquire right-of-way and provides the construction funding match on eligible safety projects. This determination is based on the locality's constraints and the potential benefits identified in the project's application.

***State and Regional Priorities***

State and regional safety priorities are articulated in the State SHSP. To engage local agencies in addressing these priorities, MDOT has created special incentives, which ties funding to projects that directly impact their goals.

MDOT has instituted an incentive program to encourage local agencies to address locations in the HSIP five percent report or other projects in line with the Michigan SHSP priority strategies. The following is a summary of the various funding practices MDOT uses to increase local participation in the State's safety program:

- Funding to local agencies for PE on safety projects at locations included in the HSIP transparency (5 percent) report, up to 10 percent of the project's design and construction costs.
- PE costs eligible for Federal funding participation for Local Safety Initiative (LSI) projects, up to a maximum of 10 percent of the project construction costs, at a rate of 50 percent Federal / 50 percent Local cost share.

- Costs associated with RSAs on Federal-participation projects are eligible for an 80 percent Federal / 20 percent Local cost share rate, as long as the local agency conducts the RSA no later than the design plans being 50 percent complete.
- PE costs associated with the optimization of traffic signal timings, eligible for Federal participation, at a rate of 80 percent Federal / 20 percent Local, with a cost cap of $5,000 per signal location.

The financial incentives developed by various scan States encourage greater participation from local roadway agencies in local safety programs.

## Summary

Funding, funding levels, and required match all play a role in local agency participation in a State's safety program. Federal and State safety funding availability encourages local agencies to advance more safety projects. However, local agencies often do not pursue these funding opportunities. Therefore, State DOTs seek to increase local agency participation in local safety programs through financial incentives and funding distribution policies.

- All scan States allocated all their HRRRP funds to local rural road safety projects
- WSDOT, IDOT, MDOT, Mn/DOT, and NJDOT all allocated varying percentages of their overall HSIP funds directly to local road safety projects.
- Mn/DOT use a data driven process based on crash rate proportional allocation to distribute their HSIP funds regardless of road ownership.
- In New Jersey, the MPOs use planning funds to assist local agencies in safety project identification and development.
- GDOT contributes required HRRRP matching funds for local agency safety projects.
- WSDOT, as well as SJTPO, fund PE for eligible local road safety projects.
- MDOT funds PE and road safety audits at pre-determined levels for safety projects tied directly to State and regional priorities, locations in the HSIP five percent report and countermeasures from the Local Safety Initiative program.

Most scan States credited financial incentive strategies with increasing local agency participation in the Federal-aid safety program. The funding distribution policies balance the needs and priorities of various stakeholders and help local agencies substantially with projects that address the highest priority issues.

## SECTION 6. TRAINING AND TECHNICAL ASSISTANCE

Local roadway safety practices vary widely from one State to another. Many local agencies are not familiar with procedures that have become standard for addressing common roadway safety issues and have received only minimal exposure to new concepts in roadway safety. Even States with successful programs require comprehensive training programs to help local practitioners develop safety solutions that move beyond basic improvements.

State DOTs have developed training and technical assistance programs to teach local practitioners and stakeholders about effective techniques for data analysis, project identification, and low-cost safety improvements. All of the scan States offer training courses through the State Local Technical Assistance Program (LTAP), Local-aid division, or affiliated university programs. The courses address best safety practices and provide the latest guidance to local agencies. States also provide varying levels of technical assistance to aid local safety practitioners with analyses and countermeasure development. This section describes some safety training and technical assistance programs offered by scan State DOTs to local agencies.

### Training

There are many local agencies within each State and many more miles of local roadways than State roadways in most States. DOTs rely on local agencies to address safety issues on non-State owned roadways. However, the average local agency engineer or planner performs multiple professional duties and may not focus significant time and attention to roadway safety. As a result, States often provide training and technical assistance to help local agencies assess critical safety issues and to develop effective projects to reduce crash incidence and severity on local roads.

### *Local Technical Assistance Programs*

Local Technical Assistance Programs were recognized as major hubs for outreach and training in all scan States. LTAPs are typically housed within State universities or a State DOT office and offer training and assistance programs to local agencies in the entire State. LTAP training courses are often available at little or no cost to participants. NJLTAP training includes courses on new guidelines, RSAs, and the use of data analysis software. The Georgia LTAP center is housed within the GDOT and offers many training programs including Road Safety 101, Road Safety Audit, Intersection Safety, Pedestrian Safety (through FHWA Resource Center) and Work Zone Safety.

The training opportunities offered by LTAPs are instrumental in providing the types of educational resources to local practitioners that they would otherwise be unlikely to receive. LTAPs teach local practitioners to follow State-supported procedures, use data to support decisions, and follow proven techniques to address local roadway safety issues.

### *Road Safety Audits/Assessments (RSAs)*

Local agencies need to identify safety issues and respective countermeasures to improve local road safety. RSAs are a highly useful tool for assessing roadway safety issues and developing potential hotspot and systemic countermeasures. Several scan States conduct RSAs on State and local roadways, dedicate resources to maintain RSA programs, and offer training on the RSA process.

The Illinois and New Jersey Departments of Transportation make extensive and flexible use of RSAs, coordinating with local agencies to identify safety issues and address them through effective countermeasures suggested during the RSA process. RSA programs are established and systematic safety reviews are done on local roadways often as part of the process to identify safety projects. In Minnesota, the counties work with individual municipalities to conduct reviews of hotspot locations or corridors. The Illinois LTAP has also created an RSA package, which includes a van, computer and materials for use when conducting an RSA. The RSA package is loaned to local agencies to support RSAs.

RSAs are an opportunity to combine multiple safety disciplines in a common purpose and ensure diverse perspectives are applied to local roadway safety. MPOs in New Jersey are actively involved in leading or participating in RSAs, which facilitates expanded involvement of participants with diverse local planning experience.

### *Law Enforcement*

Law enforcement agencies are viewed by the public as highly credible in matters of roadway safety. Law enforcement agencies can be a particularly strong partner for local transportation agencies when mutual understanding of both agencies' priorities and limitations is achieved. However, law enforcement officers may be unfamiliar with the best practices and most effective approaches to address infrastructure safety issues on local roadways.

IDOT has found that a key factor in successful HSIP and HRRRP projects is the engagement of law enforcement with State and local transportation engineers to identify safety issues and countermeasures. It is recommended that all RSA teams in Illinois include a law enforcement member. IDOT and the Illinois State Police now coordinate to provide RSA training for law enforcement officers. This training ensures that officers are exposed to RSA principles before participation on the RSA team. Law enforcement officers in Champaign-Urbana also requested and received training in the FHWA's Manual on Uniform Traffic Control Devices (MUTCD). The MUTCD training improved the officers' ability to communicate with local and State engineers about roadway designs and improvements.

The sense of shared responsibility for local roadway safety has helped to facilitate a strong partnership between IDOT and law enforcement. This initiative has resulted in greater appreciation by IDOT of enforcement capabilities and has developed more effective and targeted strategies to address local roadway safety issues in Illinois.

### *Local Public Officials*

Local agency officials have many competing demands for their limited resources. Providing them with road safety resources and training provides tools necessary to make roads safer for their citizens.

Michigan's LTAP offers a safety training program for local public officials to address misconceptions about safety programs and improve cooperation. The program features two separate courses: Common Sense Solutions for Intersections (a one day course) and What Elected Officials Need to Know about Traffic Safety (a 3.5 hour course). The program has resulted in increased support for transportation safety projects from elected officials.

### *Training Prerequisite*

Transportation safety is an evolving field. To ensure competence throughout the State, ALDOT requires all county engineers to participate in data analysis and low cost safety improvement training in order to be eligible for HRRRP funds. This requirement teaches the county engineers about approaches to identifying sites with safety issues, low-cost safety countermeasures to address them, and the eligibility and benefits associated with HRRRP projects.

## Technical Assistance

The Michigan Department of Transportation's Local Safety Initiative (LSI) supports local road authorities to reduce the number of fatal crashes and serious injuries on local roadways. Approximately 60 percent of the fatal crashes in Michigan occur on locally-owned roadways. The LSI addresses safety on the locally-owned roadway system by providing traffic engineering expertise and services directly from MDOT to local agencies. Typically, LSI staff will analyze crash data, identify candidate locations for study, and assist local practitioners in reviewing and evaluating safety improvements, consistent with the SHSP, on individual corridors. Local agencies are also trained on basic safety analysis procedures and provide recommendations for enhancements to the RoadSoft safety module. The LSI program also provides local agencies with a resource packet and support in using the RoadSoft safety module in their data analysis software.

In addition to the initial contact, the LSI establishes a "lifeline" for local agencies through their relationship with the DOT. Communication and collaboration is fostered for future safety projects. MDOT endeavors to be responsive and accessible to the agencies, while avoiding an overbearing "big brother" approach. MDOT has explicitly strived to implement an outreach program that fosters partnerships.

### *Annual Safety Summit*

Conferences and summits provide a forum for safety practitioners throughout the State to share knowledge and experience with peers and potential partners. Collaboration and information sharing is a critical component to advance local roadway safety initiatives across the State.

All scan States hold a transportation conference or summit at least once a year. In addition, several States have a transportation safety-specific conference. These gatherings often include classes, forums, and open discussion about funding, low-cost treatments, crash statistics, and other relevant topics. Participants use the summits as networking opportunities to build partnerships and encourage collaboration. Most importantly, the annual safety summits are a forum to share information with local practitioners and recognize important achievements in local transportation safety.

***Participation***

These broad-based safety meetings are generally conducted annually as a forum for local agencies to meet with State DOT representatives. The State DOTs use this annual opportunity to answer questions about the application process, respond to technical inquiries, and discuss effective countermeasures. Annual safety meetings may be co-sponsored by other State agencies, such as planning offices or other safety offices. The summits typically include a broad spectrum of transportation safety organizations, such as the American Automobile Association (AAA) and Mothers Against Drunk Driving (MADD). Given the unique opportunities afforded by a large gathering of safety practitioners at all levels of the industry, MDOT encourages county and city officials to attend by offering scholarships. The scholarships cover the full cost of registration, lodging, meals and mileage for one person from each local agency with a location on the HSIP 5 Percent High Crash Report.

***Recognition***

NJDOT uses their transportation safety conference as an opportunity to formally recognize agencies and organizations that have shown safety initiative or achieved crash abatement on local roadways. Each discipline of the transportation safety realm (engineering, enforcement, emergency response and education) are recognized for their accomplishments. Awards may also be given to safety champions. Local media are sometimes invited to attend and press releases are used to inform news outlets about the awards. This allows local agencies to receive public recognition for successful programs, and gives the State the opportunity to promote community awareness of transportation safety.

## Summary

Training programs and technical assistance are critical to developing competent local road safety practitioners and maintaining sustainable safety programs in local agencies. State DOTs along with State LTAPs have created mechanisms for training local road personnel and providing technical assistance programs.

- RSAs are a major component of several scan States' safety programs. RSAs provide a tool to identify locations with safety issues and countermeasures, but it also promotes safety awareness through the multi-disciplinary teams.
- New Jersey has a well developed RSA program that includes RSA Peer-to-Peer assistance.
- IDOT conducts RSA training for law enforcement personnel.
- To improve competence at the local level, ALDOT requires local engineers to take training courses in data analysis and low cost safety improvements in order to receive Federal funding.
- The LTAPs are the primary source for statewide roadway safety training. LTAPs train local practitioners in State-supported procedures, the use of data to support safety decisions, and to identify proven techniques to address local roadway safety issues.
- MDOT has developed courses through the LTAPs for elected officials.
- Several scan States hold annual safety summits where practitioners and stakeholders from Federal, State, and local levels to share their knowledge of effective safety programs and strategies.

The training and assistance programs discussed in this section provide States with potential avenues to improve transportation safety proficiency at the local level. Local practitioners are more likely to develop projects that reduce crashes and minimize injuries with appropriate training and technical assistance.

## SECTION 7. PARTNERSHIPS AND OUTREACH

Roadway safety is the responsibility of numerous agencies within each State. Collaboration between agencies is critical to achieve the SHSP goals to reduce crashes on all public roads. Active communication is required between Federal, State, regional and local government to improve safety. In many States, the DOT is usually charged with the coordination between these agencies.

State DOTs must also find ways to reach out to potential partners to encourage communication and enable all stakeholders to most effectively contribute to the State's ultimate safety goals. Working as a team, through carefully cultivated partnerships, can eliminate redundant efforts, increase program efficiency, and effectively use limited financial and personnel resources. Coordination and cooperation can be beneficial at any stage of project or program development.

## Partnerships

Many of the successes described by the scan States has been through the multi-agency collaboration and organizations which aligned priorities around the goals of the SHSP. Collaboration between different organizations, including engineers, planners, educators, community leaders, activists, law enforcement, and emergency medical services can bring diverse safety expertise to existing programs and produce ideas for innovative approaches. These partnerships are opportunities to share available resources, provide training, and develop innovative perspectives on solutions and countermeasures.

***Local Coalitions and Committees***

When local agencies and organizations are able to collaborate, they can develop a collective understanding of the major safety issues and develop the best approaches to prevent or address safety issues.

In Minnesota, local agencies have formed multi-disciplinary Toward Zero Deaths (TZD) coalitions to address traffic safety issues in their communities. These coalitions address both infrastructure and behavioral safety issues in their communities. The TZD program has resulted in statewide safety laws and outreach programs to address some of the root causes of crashes or injury (e.g., drunk-driving, seat-belt usage, texting, and teen driver inexperience). The TZD program also contributed to the creation of additional hospital trauma centers to treat crash victims and reduce fatal crashes.

Minnesota has also encouraged and fostered partnerships and collaboration at the county level through the formation of the County Highway Safety Committee. The Minnesota County Highway Safety Committee meets regularly and has actively lobbied for key safety laws and funding. The Committee facilitated multi-agency projects that produced better unit costs and an increased scope of improvements on more local roadways than standalone projects.

### *Metropolitan Planning Organizations*

Effective responses to common safety issues require strong and collaborative relationships between State, regional, and local transportation agencies. Leadership from MPOs can encourage better data analysis, identify projects that reflect regional priorities, and enhance collaboration between stakeholders on local roadway safety programs.

The Southeast Michigan Council of Governments (SEMCOG), the MPO for counties in the Detroit metropolitan area, has taken a very active role in the region's highway safety efforts. SEMCOG has adopted comprehensive approaches to safety planning by utilizing transportation data to forecast and mitigate potential hazards. They have integrated transportation safety into a larger planning context to address shifting traffic, pedestrian, bicycle, and public transportation needs in their communities. SEMCOG maintains and analyzes regional crash data using GIS technology and provides the results to the local member agencies and the public via web- based applications. The web-based crash analysis tool allows local agencies to create a list of the intersections with the highest number of crashes (within 150 feet). SEMCOG organizes and conducts RSAs on local roadways and pursues other local transportation safety efforts such as deer crash abatement and elderly mobility safety improvements. SEMCOG's programs fulfill a need for its communities by consistently addressing shared issues and developing systemic solutions to minimize crashes.

The Delaware Valley Regional Planning Commission (DVRPC), a bi-State MPO for the Philadelphia/Trenton metropolitan area, established a multidisciplinary Regional Safety Task Force. The task force is comprised of safety professionals from engineering, education, law enforcement, emergency services, and legislators in two States. The task force meets quarterly to exchange information on safety strategies and resources that can reduce crashes and injuries in the region. With guidance from the task force, DVRPC developed a Regional Safety Action Plan to reduce crashes by focusing support on the following efforts:

- Curb aggressive driving.
- Reduce impaired driving.
- Keep vehicles on the roadway.
- Sustain safe senior mobility.
- Increase seat belt usage.
- Improve the design and operation of intersections.
- Ensure pedestrian safety.

Through the Regional Safety Action Plan, DVRPC works to educate local engineers to identify potential safety issues, select effective countermeasures, and develop methods to evaluate the cost-effectiveness of countermeasures. One of the task force's principal goals is to foster partnerships between safety stakeholders to improve safety in the bi-State region. DVRPC also offers training resources, such as free workshops, for local stakeholders on safety topics.

The Illinois Champaign-Urbana MPO identified a need for a safety specialist within their organization. They partnered with another strong safety advocate, State Farm Insurance, to fund the position as a pilot program. Through this position, the MPO expanded its ability to conduct RSAs and to promote safety concepts within the region. This special partnership helped to develop new local roadway safety projects.

***Law Enforcement***

Law enforcement is an active safety partner in a number of the scan States. This partnership lends additional credibility to local roadway safety projects. The relationship between law enforcement and local transportation agencies is very important in Illinois, where law enforcement partners are able to elevate safety awareness at a local level with an authority that may not be possible for the State DOT. IDOT and the State Police have organized quarterly meetings to discuss goals, areas of concern, and current actions.

In New Jersey, relationships between NJDOT and law enforcement has been fostered through the common goal of reducing fatalities on the State's roadways. Typically, the partnership is focused around the enforcement and educational aspects of transportation safety for local agencies and communities. New Jersey also reported that the collaboration has come to include active engagement of law enforcement in road safety audits. This collaboration develops more effective countermeasures to address local roadway safety issues because often times the officers bring information to the table that is not evident in the crash reports.

All of the partnerships discussed in this section are cultivated by active communication and investment of time by the State DOTs and other safety stakeholders.

## Outreach

To successfully address local roadway safety, State DOTs need to find cohesive and unobtrusive ways to build relationships and trust with local transportation agencies. The scan States are building relationships through outreach activities such as conferences, individual interaction with local engineers, and regional education efforts.

### *Conference Presentations*

State DOTs need to communicate with county and municipal engineers to increase local participation in and understanding of local safety programs. Heavy workloads and busy schedules force State DOTs to concentrate on forums that can maximize the delivery of information. Local engineering conferences offer an opportunity to present safety program information to local road practitioners.

Alabama has a very active County Engineers Association. ALDOT uses the Association's annual meetings to conduct training and information-sharing sessions. ALDOT presents critical information, which may include changes to application procedures, new statewide priorities, or the types of safety projects most likely to receive approval, at these meetings. The county engineers have cited the meetings and the overall collaboration between ALDOT, the FHWA Division Office, and local agencies as key to developing successful projects.

The County Road Association of Michigan (CRAM) holds an annual County Engineers' Workshop. MDOT representatives typically participate in this workshop, as part of an outreach effort to discuss safety issues related to new and upcoming projects, effective countermeasures, local agency program applications, and other related topics. The workshop is not solely focused on safety, but transportation safety is one of the primary areas of discussion. This workshop fosters continued collaboration between the State and county officials and results in improved Federal safety project applications from county engineers.

### *State-Local Liaisons*

It can be difficult for State DOTs to find effective ways to provide continuous and direct assistance to local engineers given the myriad of local transportation agencies spread throughout a State. Many States are unsure how to engage local agencies in the overall safety program process.

GDOT created Off-System Safety Coordinator positions within each of the seven field districts to manage their local road safety program. The program, referred to as Off-System Safety Program, works with local governments to provide safety improvements on local roadways. Off-System Safety Coordinators are consultant positions, primarily comprised of retired staff members from GDOT, with extensive experience in working with the local agencies. Individuals chosen to fill these positions often have previously established long-standing working relationships with county and municipal engineers. This improves their ability to respond to inquiries and work effectively with policymakers.

The established relationships between the Off-System Coordinators and the local agencies are a valuable tool to engage local agency participation in the HRRRP application process. GDOT solicits interest from the local agencies, which then request assistance from the Off-System Coordinators and GDOT District Engineers to address safety concerns. Coordinators and GDOT engineers then meet with the local agencies and conduct field assessments to identify safety needs and countermeasures. The Off-System Coordinators assist the local agencies by providing technical knowledge and traffic engineering expertise to identify safety problems, develop countermeasures, and prepare cost estimates.

The coordinators in this program provide local agencies with oversight and assistance with Federal-aid requirements. For the local agencies that do not have engineering staff, the assistance from the State representatives is critical to implement safety countermeasures. The mentoring provided by the liaisons has developed champions for safety in the participating local agencies.

### *MPO Committees*

Local agencies may have difficulty coordinating across jurisdictional boundaries. Regional planning agencies can serve as a bridge to work with a broad array of safety stakeholders in a region and coordinate across jurisdictional boundaries. MPOs can also effectively educate the public about safety issues in the region.

The South Jersey Transportation Planning Organization (SJTPO) created a Community Traffic Safety Program (CTSP) – the South Jersey Traffic Safety Alliance (SJTSA) – to address safety needs in their four-county region. The Alliance brings together traffic safety professionals from the fields of law enforcement, education, fire, rescue, engineering and planning to develop region-wide traffic safety programs, share successful practices, exchange information, and support capital projects. SJTPO staffs the SJTSA with one full-time program manager and three part-time transportation safety specialists using metropolitan planning (PL) funds.

The Alliance conducts outreach efforts to improve citizen awareness of transportation safety issues and steps they can take to improve safety in the region. Some of these programs include the following:

- Buckle Up Stencil, Saved By The Seat Belt Club, and Seat Belt Surveys – outreach programs to educate drivers about the benefits of seat belt use and encourage greater seat belt use by drivers and passengers in motor vehicles.
- Bounty Program – outreach education program to educate the public about the proper disposal for old and used car seats as well as the danger of reusing these seats.
- Defensive Driving – outreach program to improve driver skills and driving techniques to prevent crashes and violations.
- Awards Program – outreach program to publicize the achievement of transportation safety champions, chosen from nominations from their peers, in the fields of education, engineering, planning, enforcement, and community awareness.

These programs focus on preventative measures that reduce both crashes and crash severity on local roadways. The MPO brings together local transportation agencies to coordinate and conduct educational programs across the SJTPO region.

## Summary

Partnerships and outreach efforts practiced by the scan States usually involve collaboration between multi-disciplinary groups, including engineering, planning, education, law enforcement, emergency services, and political representatives.

- Metropolitan Planning Organizations are safety champions who provide coordination and public outreach to local agencies in New Jersey, Michigan and Illinois.
- The Minnesota County Engineer Safety Association is uniquely positioned in the State to impact local road safety policies and funding. As a collective group they are able to lobby for policies and programs that address their roadway safety and share information among local agencies.

- State DOTs use County Engineer Safety Association meetings and workshops to present information about the Federal-aid process. The meetings or workshops clarify statewide priorities, expose a large group of local engineers to the Federal-aid process, and improve the quality of project developed by county engineers.
- Minnesota's Toward Zero Deaths community coalitions develop safety awareness at the local level through collaboration of multi-discipline stakeholders addressing safety in their communities.
- IDOT and NJDOT nurture relationships with law enforcement through the common goal of both groups to save lives on the roadway network.
- GDOT leverages the relationships of retired employees with local agencies by bringing them back as consultants to work closely with local agencies on safety issues.

## SECTION 8. LOCAL AGENCY SUCCESS STORIES

At each scan visit the team was presented with local agency safety initiatives that demonstrated a level of success from their local safety programs/projects. The success stories provided by each of the local agencies offer examples of interesting and noteworthy approaches or programs that some of the scan States have used to improve safety on local roadways. Many of the States' stories highlighted programs that achieved significant improvements in analyzing, identifying, and addressing crash issues on local roadways.

This section presents only some of these initiatives that advance local agencies' safety awareness and commitment.

### Georgia Local Success Story: Douglas County Safety Action Plan

Douglas County, Georgia created a County Curve Action Plan as part of the Georgia Department of Transportation's Safety Action Plan Program. This plan has been particularly helpful to the county as it supported funding opportunities through the Georgia Off-System Safety Program.

One of the principal components of their safety action plan, which aligns with the Georgia SHSP, is a focus on roadway departure crashes along curves. Horizontal curves on local and rural roadways represent a major concern in Douglas County, because many of their roads are former wagon trails that were paved over time without addressing the roadway alignment, shoulders, clear zones, and lighting to meet current standards. Consequently, the County experiences a significant number of run-off-road (ROR) crashes associated with these curves.

With a lack of comprehensive data, the County depended on their County Road Department staff, historic knowledge from the sheriff department, and residents to supplement the available data to develop the Curve Action Plan. High-risk locations were identified. A consultant was hired to conduct a qualitative analysis of critical locations and identify countermeasure strategies where appropriate. Improvement strategies included signing and striping modifications, as shown in Figure 4, as well as centerline and edgeline raised pavement markers.

Photograph: Douglas County Road Department.

Figure 4. Curve Ahead Pavement Markings and Signage.

There has been a significant reduction in ROR crashes since the county began implementing low cost strategies such as warning signs, chevron signs and raised pavement markers along the curves as part of their "Curve Safety Plan". After installing the double advance warning signs, chevrons, and raised pavement markers, a number of crashes continued to occur along some curves. The County took another step by adding the Pennsylvania DOT "curve ahead" markings on the pavement, including arrows. The pavement markings have made a tremendous difference as there have been no crashes in the treated locations since the installation.

## Minnesota Local Success Story: Township Sign Inventory and Replacement Pilot

The Township Sign Inventory and Replacement pilot program was developed by the Minnesota Association of Townships (MAT). This program was funded by HSIP and State funds and administered by the State Aid for Local Transportation (SALT) division of Mn/DOT in coordination with MAT. This pilot program began with the townships located within six counties and was broken into two phases, Engineering and Construction.

The goal is to develop and refine the requirements for sign reduction and removal to assist agencies in complying with new retroreflectivity standards. MAT has set a 10 to 15 percent reduction goal in total sign count. The first phase was an engineering assessment focused on the evaluation of existing signs to develop a sign inventory, a set of sign plans, and provisions for each county broken out by township. Mn/DOT used a consultant to collect data, including an inventory and assessment of sign posts. The second phase involved the construction or installation of signage improvements, based on the results of the engineering assessment. Based on the experience of this pilot, Mn/DOT is developing a Sign Reduction Manual to help local agencies comply with MUTCD guidelines. The manual will also address when it is permissible and appropriate to remove a sign that is no longer of value.

Funding for the program was acquired when the Township Association of MN formed a partnership with State, county and local industry and consultants interested in transportation safety. These transportation partners traveled to Washington, D.C. to solicit congressional support for Federal funding for the sign replacement plan and proposed safety-related benefits. Two years after the Minnesota Association of Townships received a $3 million Federal earmark under SAFETEA LU, the townships were granted an additional $2.5 million in State contributions to match the Federal funds and continue the sign replacement program. The State match was seen as a critical step to completing the pilot program given limited local funding. The program started with six pilot counties and a limit of two years has been established to spend the Federal funds.

This program is helping townships to establish consistent signage on all corridors. In some cases, individual townships overcame coordination difficulties by taking the initiative and sharing responsibilities with other local agencies based on areas of expertise. No formal evaluation of the project has been completed yet, but there is potential to conduct research on driver behavior impacts in the future. MAT has selected townships in two counties for the next round of this project. SALT has begun the initial inventory and plan development phase.

## Washington Local Success Story: Pierce County Safety Program

With guidance from the Washington DOT, Pierce County has adopted a systematic approach for the local safety improvement program. The program includes a prioritization process that is data-driven and focuses on the most severe crash types and hotspot locations, as sanctioned both by WSDOT and other local agencies.

Pierce County has found that systematic field inventories and assessments of existing infrastructure have improved opportunities to identify safety issues on county roadways. The county conducts the following efforts to proactively identify potential projects:

- Inventories for all pavement markings and scattered inventories of traffic signal equipment and operational attributes.
- Dedication to preventative maintenance, including frequent sign inspections and traffic signal maintenance (e.g., checking traffic signals every other month).
- An aggressive daytime traffic sign inspection program has been augmented with a new nighttime program.

Pierce County standardized its evaluation procedures and emphasizes documentation to support the HSIP applications, improve the potential for Federal funding, and guide eventual project implementation. To maintain consistency, the County developed a standard format for traffic studies and based their traffic safety device policies on MUTCD guidance, which has reduced potential political conflicts. Pierce County requires road crews to thoroughly document field assessments, maintenance projects, and roadway modifications.

Through the recently adopted approach, Pierce County developed county safety projects for roadside tree removal and installing the roadway safety edge. The systematic approach used to obtain funding for their current HSIP projects has resulted in funding to remove approximately 300 roadside trees, which were selected during the project development based on existing or potential safety concerns. Prior to this project, Pierce County completed other systemic safety upgrades on county roadways including guardrail upgrades and flashing beacon warnings at intersections. All of these projects address high-priority SHSP emphasis areas.

## New Jersey Local Success Story: NJTPA Regional Safety Priorities Program

The North Jersey Transportation Planning Authority (NJTPA) has implemented a Regional Safety Priorities Program that takes a data-driven systematic approach to identifying projects. Their success can be seen in the number of projects that they have successfully initiated and completed and the total amount of funding they have received and utilized.

NJTPA is committed to incorporating safety conscious planning principles in their planning processes. In 2005, the NJTPA undertook a collaborative effort with local agencies in the region to develop safety priorities for their Regional Transportation Plan. A consultant was employed to help identify the most pressing safety needs in the region. The multi-modal Regional Safety Priorities study developed a comprehensive set of new safety initiatives which included engineering, enforcement, and educational countermeasures. The Regional Safety Priorities Update, completed in 2008, built upon the results of this study. Both studies were forward-looking, multi-modal efforts to integrate safety considerations into all phases of transportation improvement planning and development, and to elevate safety to a high priority at all levels of decision-making.

In addition to the safety initiatives, the studies identified about 25 priority locations in the region with a history of crashes or safety issues related to vehicles, pedestrians, transit, bicycles, trucks, and elderly users. In order to maximize the potential benefits, NJTPA used regional crash data to identify critical locations, such as those demonstrating left-turn, pedestrian, and cross-over crashes. Crash data was geo-coded and high-crash locations were identified. NJTPA conducted RSAs, bringing together engineers, law enforcement personnel, planners, elected officials and stakeholders at each location to identify critical safety issues and suggest improvements.

NJTPA concentrated on short-term and relatively low-cost improvements such as signing, improvements to existing signals, and pavement markings for pedestrians, that could be implemented by local agencies and funded through the NJTPA Local Safety Program, and to the extent possible, the HRRRP. Over 75 percent of locations and needs identified in the first study have been addressed through engineering, enforcement and educational strategies, and many through the Local Safety Program.

# SECTION 9. OBSERVATIONS AND CONCLUSIONS

The diversity among local jurisdictions and governments in each State can make the development and implementation of a standardized local road safety program difficult. Some States have worked through the challenges of integrating local roadways in their safety programs and adopted practices to assist them. The Local Road Safety Domestic Scan identified several of these practices that if adopted can help State and local agencies develop programs to meet their specific needs. Some practices identified on the domestic scan and documented in this report may not be easily adopted by other States because they may be based on particular State legislative policies. However, most of the practices are readily adoptable or adaptable to many States, with some coordination and collaboration with safety stakeholders.

States demonstrating reductions in local roadway crashes have depended on the integration of a variety of safety programs and funding mechanisms at the local level. Success relies first on a comprehensive, coordinated and well-planned approach that is data-driven and results oriented. States and local agencies must also identify procedures to improve local participation and target resources and actions where they will do the most good. Many of the most productive local road safety programs depend on leadership from the State DOT and partnerships between many safety stakeholders.

## Primary Observations from the Domestic Scan

### *Data Collection and Analysis*

Access to crash data to support a data-driven decision making process is an integral part of getting local roadway involvement in safety improvement programs. The HSIP and other safety programs depend on the results of reliable crash data and analyses to direct funding to the most effective safety projects. In an effort to build a comprehensive crash data system that includes local road data, IDOT allocated $1 million of HSIP funds to county governments to collect and geo-code their roadway crash data. All scan States have data analysis software tools. NJDOT, ALDOT, Mn/DOT, MDOT and GDOT make these tools and respective user training accessible to local agencies. IDOT and WSDOT conduct comprehensive data analysis and share the results with local agencies. This has proven effective in encouraging local road practitioners to take an active role in identifying safety issues and seeking resources to address them.

In Georgia and Illinois, regional planning agencies provide technical assistance for member agencies and align project identification with regional safety priorities. Direct technical assistance helps local agencies identify specific safety issues, develop safety countermeasures, and apply for HSIP or other safety funding.

### *Local Project Selection*

In many cases, local safety programs are structured to guide local agencies to identify safety projects that directly address contributing infrastructure factors associated with the most severe crashes. For example, planning tools, such as the SHSP and local roadway safety plans, guide local agencies in their project selection. Mn/DOT utilizes visual reference tools to guide local agencies in their project selection process. Local agencies typically identify priority locations for funding and implementation; however, priority is usually given to those projects that address the State's safety goals. MDOT gives priority to locations listed in their HSIP five percent report for funding. For MDOT and Mn/DOT system-wide strategies are favored for safety funding. The project identification practices used in the scan States provide resources for local transportation agencies to effectively pursue State and Federal transportation funding.

### *Local Project Administration*

State DOTs with Local-aid divisions within their organizational structure have dedicated personnel to concentrate their efforts on local road issues. Local-aid divisions are instrumental in helping local agencies navigate the State and Federal-aid processes and provide technical assistance. Dedicated State DOT Local-aid divisions enhance local agency participation in safety programs. These divisions understand the various funding mechanisms for local safety projects and provide essential support for local agencies to acquire and obligate funding. All scan States boast Local-aid divisions. The Local-aid divisions also offer safety-related technical assistance either through dedicated local safety staff or the Safety Office within the DOT.

Including local agencies in the development of methodology and guidelines for State and Federal-aid funding fosters a local agency buy-in for safety programs in Alabama. Scan States have employed a variety of time saving procedures to expedite safety project implementation at the local level. MDOT and ALDOT use force accounts where the local agencies provide labor for safety projects. NJDOT requires projects meet categorical exclusion criteria to be eligible for safety funding. The scan State Local-aid divisions are representative of effective administration structures and practices that may serve as models for other States.

### *Funding*

Availability and access to Federal and State safety funding is instrumental to the integration of local agencies in the States' safety programs. States distribute funds to local projects via State DOT districts or regional MPOs. States must balance the safety needs and priorities of various stakeholders. All scan States distribute 100 percent of HRRRP funds to local roads. Additionally, varying percentage of their overall HSIP funds are dedicated to improving safety on local roadways. In addition, some States offer financial incentives to encourage local agency participation in safety programs. These financial incentives include funding local road safety projects 100 percent and often providing PE costs.

### *Training and Technical Assistance*

Providing roadway safety-related training and technical assistance to local practitioners promotes safety awareness and builds a safety culture at the local roadway level. Safety training and technical assistance provides practitioners with the knowledge and technical capabilities to succeed in local roadway safety programs. In the scan States, LTAP Centers provided free or low-cost roadway safety training to support States' safety programs. Transportation safety training and assistance programs improve proficiency among local practitioners involved in project development. Highly competent local practitioners are more likely to develop projects that reduce crashes and minimize severity. ALDOT requires county engineers the complete data analysis and low cost safety improvement training to be eligible for safety funds. MDOT's Local Safety Initiative provides hands-on technical assistance to local practitioners. Training for elected officials and law enforcement personnel encourages participation from these stakeholders in safety projects and improves efficiency and collaboration.

***Outreach and Partnership***

Strong partnerships between the State, regional, and local agencies enable peers to share critical safety information. Many of the scan States have strengthened transportation safety collaboration between the 4Es. These partnerships include local agency coalitions, such as Minnesota's Towards Zero Death coalition, which have helped pass safety legislation related to reducing drunk driving, seat belt violations, teen driver crashes and fatal crashes. Partnerships between local agencies can lead to multi-agency projects with broad scopes in multiple jurisdictions at reduced construction costs. As shown in Illinois, partnerships between engineers and law enforcement have been useful in identifying safety issues and countermeasures to address them.

MPOs are uniquely positioned to address regionally significant safety issues and provide support to member agencies. In New Jersey, MPOs administer the local road safety funding through the regional TIP. MPOs provide significant technical assistance to local agencies, such as identifying projects through data analysis or leading RSA programs to develop safety treatments. MPOs are also well positioned to develop educational outreach programs that promote preventative measures to reduce both total crashes and severity on local roadways.

To elevate local roadway safety programs and increase public awareness, States must find ways to effectively communicate and reach out to local agencies and other organizations. New Jersey and Michigan hold annual safety summits to exchange safety information and publicly recognize specific accomplishments in roadway safety. Several States have promoted safety and built partnerships through their technical assistance programs.

***Conclusion***

Understanding the successful practices experienced in the scan States will provide other States with the impetus to try new or modify existing strategies that best fit their own local roadway safety programs. States and local agencies can use these documented practices as a model to improve data collection and analysis procedures; initiate new application or funding mechanisms; streamline existing safety programs; institute strong technical assistance programs; or develop valuable partnerships with other stakeholders. The means exist to significantly improve local roadway safety in nearly every State. The results of the Local Road Safety Domestic Scan provide a resource for all transportation safety practitioners to use in developing or improving local roadway safety programs.

# APPENDICES

## State Fact Sheets

The following pages contain State briefing sheets that contain information gathered prior to the scan. The information was generally used to provide basic context, regarding the State structure and safety programs, for the Scan Team members visiting each State.

## Alabama

### *State Overview*

Alabama covers an area of 54,419 square miles in the Southeastern region of the United States. According to the United States Census Bureau their population in 2009 was estimated at 4,708,708. The State has approximately 96,045 miles of public roads; close to 87 percent are locally owned and maintained.

### *Government Structure Related to Transportation*

The State of Alabama is comprised of the following government offices:

- 13 Metropolitan Planning Organizations
- 9 Divisions
- 67 Counties

According to the State website, the Department of Transportation (ALDOT) is organized into nine geographic regions called Divisions, with a Central Office located in Montgomery. The Central Office is organized into the Office of the Transportation Director and the Office of the Chief Engineer with Bureaus and Divisions reporting to the Chief Engineer. Additionally, the Department has several other boards and committees that operate either within a bureau or as a cooperative effort among several bureaus and/or divisions.

The ALDOT County Transportation Bureau (CTB) has dedicated personnel for local roads and serves as the liaison between the Alabama Department of Transportation and the 67 counties in Alabama. The CTB has about 15 engineers in the Central Office, one engineer in each District, and about three support staff members.

### *Federal Safety Funding*

Alabama received approximately $30.9 million in Federal safety funding in 2009. Funding is divided between the Highway Safety Improvement Program (HSIP) and High Risk Rural Roads Program (HRRRP) as shown in the table below. The division office of the bureau prescreens local safety projects, provides construction engineering services to the local agencies, and oversees Federal-aid projects.

| HSIP | HRRRP |
|---|---|
| $24,940,307 | $1,810,055 |

### *Safety Data*

The State's two year average of fatal crashes, using FARS data for 2007 and 2008, was 522 fatal crashes on State roads and 427 fatal crashes on locally owned and maintained roads. As shown in the following table, approximately 45 percent of fatal crashes occurred on locally maintained roads. Route signing was used to determine roadway classification.

| Federal/State Roadways | Local Roadways |
|---|---|
| 55% | 45% |

## Georgia

### *State Overview*

Georgia covers 59,425 square miles in the Southeastern region of the United States. According to the United States Census Bureau their population in 2009 was estimated at 9,829,211. The State has approximately 117,645 miles of public roads. Approximately 84 percent of the State's roadways are locally owned and maintained.

### *Government Structure Related to Transportation*

The State of Georgia is comprised of the following government offices:

- 15 Metropolitan Planning Organizations
- 6 Districts
- 159 Counties

The Georgia Department of Transportation (GDOT) Off-System Safety Program was established as part of the GDOT Safety Action Plan to take assertive action in reducing the number of fatal crashes on off-system or local roadways.

The ALDOT County Transportation Bureau (CTB) has dedicated personnel for local roads and serves as the liaison between the Alabama Department of Transportation and the 67 counties in Alabama. The CTB has about 15 engineers in the Central Office, one engineer in each District, and about three support staff members.

### *Federal Safety Funding*

Georgia received approximately $55 million in Federal safety funding in 2009. Funding is divided between the Highway Safety Improvement Program (HSIP) and High Risk Rural Roads Program (HRRRP) as shown in the table below. The Federal funding from the HRRRP is administered through the Off-System Program efforts for distribution to counties and municipalities.

| HSIP | HRRRP |
|---|---|
| $52,015,784 | $3,149,726 |

### *Safety Data*

The State's two year average of fatal crashes, using FARS data for 2007 and 2008, was 850 fatal crashes on State roads and 581 fatal crashes on locally owned and maintained roads. As shown in the following table, approximately 41 percent of fatal crashes occurred on locally maintained roads. Route signing was used to determine roadway classification.

| Federal/State Roadways | Local Roadways |
|---|---|
| 59% | 41% |

## Illinois

### *State Overview*

Illinois covers an area of 57,918 square miles in the Midwest region of the United States. The United States Census Bureau estimated the State's population was 12,910,409 in 2009. The State has approximately 138,833 miles of public roads. Approximately 88 percent of the State's roadways are locally owned and maintained.

### *Government Structure Related to Transportation*

The Illinois Department of Transportation's (IDOT) Bureau of Local Roads and Streets has about 25 full-time employees in its central office. Approximately 100 IDOT staff members work on local roadway programs. This includes IDOT District Local Roads and Streets personnel.

The State of Illinois is comprised of the following government offices:

- 13 Metropolitan Planning Organizations
- 9 Districts
- 102 Counties
- 2,000 Municipalities/Townships

### *Federal Safety Funding*

Illinois received approximately $45.5 million in Federal safety funding in 2009. Funding is divided between the Highway Safety Improvement Program (HSIP) and High Risk Rural Roads Program (HRRRP) as shown in the table below. The Federal funding is administered through the IDOT's Bureau of Local Roads and Streets.

| HSIP | HRRRP |
|---|---|
| $42,434,685 | $3,024,273 |

### *Safety Data*

The State's two year average of fatal crashes, using FARS data for 2007 and 2008, was 614 fatal crashes on State roads and 425 fatal crashes on locally owned and maintained roads. As shown in the following table, approximately 41 percent of fatal crashes occurred on locally maintained roads. Route signing was used to determine roadway classification.

| Federal/State Roadways | Local Roadways |
|---|---|
| 9% | 41% |

## Michigan

### *State Overview*

Michigan covers 97,990 square miles and is located in the Great Lakes region of the United States. According to the United States Census Bureau their population in 2009 was estimated at 9,969,727. Michigan is the largest State east of the Mississippi River by total area. The State has approximately 121,456 miles of public roads. Approximately 90 percent of the State's roadways are locally owned and maintained.

### *Government Structure Related to Transportation*

The Michigan Department of Transportation (MDOT) has two full time employees dedicated to safety. One of the keys to success of the local safety program is MDOT's Local Safety Initiative.

The State of Michigan is comprised of the following government offices:

- 12 Metropolitan Planning Organizations
- 83 Counties
- 400+ Cities/Villages
- 520 Local Agencies
- 13 Rural Task Forces

### *Federal Safety Funding*

Michigan received approximately $42 million in Federal safety funding in 2009. Funding is divided between the Highway Safety Improvement Program (HSIP) and High Risk Rural Roads Program (HRRRP) as shown in the table below. Michigan Department of Transportation's Local Safety Initiative program administers HSIP and Local Rural Road funds to local agencies using an "Annual Call" for projects.

| HSIP | HRRRP |
|---|---|
| $39,131,608 | $2,926,006 |

### *Safety Data*

The State's two year average of fatal crashes, using FARS data for 2007 and 2008, was 385 fatal crashes on State roads and 568 fatal crashes on locally owned and maintained roads. As shown in the following table, approximately 60 percent of fatal crashes occurred on locally maintained roads. Route signing was used to determine roadway classification.

In order to remain a recognized leader in improving safety on roadways, the State of Michigan Strategic Highway Safety Plan (SHSP) has set two ambitious safety goals:

| Federal/State Roadways | Local Roadways |
|---|---|
| 40% | 60% |

- Reduce fatal crashes to fewer than 850 by 2012.
- Reduce serious crash injuries from 7,485 in 2007 to fewer than 5,900 by 2012.

## Minnesota

### *State Overview*

Minnesota covers 86,943 square miles in the Midwest region of the United States. According to the United States Census Bureau their population in 2009 was estimated at 5,266,214 and 60% of the population resides in the Minneapolis - St. Paul metropolitan area. The State has approximately 132,048 miles of public roads. Approximately 89 percent of the State's roadways are locally owned and maintained.

### *Government Structure Related to Transportation*

The State of Minnesota has the following jurisdictions/agencies at the local level:

- 8 Districts
- 8 Area Transportation Partnerships (ATPs)
- 87 Counties
- 854 Cities
- 1,786 Organized Townships

The Minnesota Department of Transportation (Mn/DOT) is headquartered in St. Paul, MN.

### *Federal Safety Funding*

Minnesota received approximately $26.8 million in Federal safety funding in 2009. Funding is divided between the Highway Safety Improvement Program (HSIP) and High Risk Rural Roads Program (HRRRP) as shown in the table below. The Mn/DOT State Aid for Local Transportation (SALT) administers funding allocations to local agencies.

| HSIP | HRRRP |
|---|---|
| $24,940,307 | $1,810,055 |

### *Safety Data*

The State's two year average of fatal crashes, using FARS data for 2007 and 2008, was 224 fatal crashes on State roads and 219 fatal crashes on locally owned and maintained roads. As shown in the following table, approximately 49 percent of fatal crashes occurred on locally maintained roads. Route signing was used to determine roadway classification.

| Federal/State Roadways | Local Roadways |
|---|---|
| 51% | 49% |

## New Jersey

### *State Overview*

New Jersey is located in the Northeast region of the United States and covers 8,729 square miles. According to the United States Census Bureau their population in 2009 was estimated at 8,707,739 and is the highest population density of any of the United States. The State has approximately 38,552 miles of public roads. Approximately 93 percent of the State's roadways are locally owned and maintained.

### *Government Structure Related to Transportation*

The State of New Jersey is comprised of the following government offices:

- 3 Metropolitan Planning Organizations
- 4 District Offices
- 21 Counties
- 566 Municipalities

One of the keys to success of the local safety program is the integrated relationship between the New Jersey Department of Transportation (NJDOT) and the regional Metropolitan Planning Organizations (MPOs). NJDOT has a separate division, the Division of Local Aid and Economic Development, which works with county and municipal government officials to improve the efficiency and effectiveness of the State's transportation system.

### *Federal Safety Funding*

New Jersey received approximately $25 million in Federal safety funding in 2009.

Funding is divided between the Highway Safety Improvement Program (HSIP) and High Risk Rural Roads Program (HRRRP) as shown in the table below.

The Division of Local Aid and Economic Development administers HSIP and Local Rural Road funds to local agencies primarily through MPOs.

| HSIP | HRRRP |
|---|---|
| $23,356,592 | $1,666,605 |

### *Safety Data*

The State's two year average of fatal crashes, using FARS data for 2007 and 2008, was 270 fatal crashes on State roads and 344 fatal crashes on locally owned and maintained roads.

As shown in the following table, approximately 56 percent of fatal crashes occurred on locally maintained roads. Route signing was used to determine roadway classification.

| Federal/State Roadways | Local Roadways |
|---|---|
| 44% | 56% |

## Washington

### *State Overview*

Washington State covers an area of 71,342 square miles in the upper northwest corner of the United States. According to the United States Census Bureau their population in 2009 was estimated at 6,664,195. Sixty percent of the State's population resides in the Seattle metropolitan area. The State has approximately 83,381 miles of public roads. Approximately 81 percent of the State's roadways are locally owned and maintained.

### *Government Structure Related to Transportation*

The State of Washington is comprised of the following government offices:

- 10 Metropolitan Planning Organizations
- 14 Regional Transportation Planning Organizations
- 39 Counties
- 200+ Cities/Villages

WSDOT has approximately 7,200 full-time employees, including engineers, vessel captains, maintenance technicians, environmental specialists, planners, and many others. WSDOT has a well-defined local roads program. The WSDOT local roads program employs 30 full-time employees in the headquarters office and 30 in the field distributed among 6 regional offices. Local roads staff includes bicycle/pedestrian specialists, safe routes to school specialists, and bridge specialists among others.

### *Federal Safety Funding*

Washington received approximately $19.6 million in Federal safety funding in 2009. Funding is divided between the Highway Safety Improvement Program (HSIP) and High Risk Rural Roads Program (HRRRP) as shown in the table below. The Federal funding is administered through Washington Department of Transportation (WSDOT) for distribution to all public road agencies.

| HSIP | HRRRP |
|---|---|
| $18,071,911 | $1,572,286 |

***Safety Data***

The State's two year average of fatal crashes, using FARS data for 2007 and 2008, was 230 fatal crashes on State roads and 276 fatal crashes on locally owned and maintained roads. As shown in the following table, approximately 54 percent of fatal crashes occurred on locally maintained roads. Route signing was used to determine roadway classification.

| Federal/State Roadways | Local Roadways |
| --- | --- |
| 46% | 54% |

## Key Contacts in Host States

This report would not be possible without support from our host States. Host States provided preliminary information prior to the visit, organized participants for the scheduled scan date, and supplied subsequent follow up information. The key participants in each State are presented below.

***Alabama***

The Scan Team met with or talked to several representatives from the Alabama Department of Transportation (ALDOT) including:

- Waymon Benifield, Safety Management Section Administrator
- Tim Barnett, Safety Operations Engineer
- Eric Marable, Safety Support Staff
- Wes Elrod, Assistant Chief, Multimodal Bureau
- Sonya Baker, Safety Support Staff
- Sam Poole, Safety Support Staff
- Ed Phillips, Chief, County Transportation Bureau
- Brad Lindsey, P.S. & E. Engineer, County Transportation Bureau
- Walter (Richie) Beyer, IV, Elmore County Engineer
- William L. Wallace, Safety Operations Manager
- Dr. Dan Turner, Professor, University of Alabama, Tuscaloosa

Additionally, the Scan Team met with Linda Guin, FHWA Safety & Technology Engineer, Alabama Division.

### *Georgia*

The Scan Team met with several representatives from the Georgia Department of Transportation (GDOT) including:

- Norm Cressman, Safety Program Manager
- Steve Sanders, District Traffic Operations Manager
- Charity Belford, Traffic Operations
- Beverly Fontenot, Staff Development/Training Coordinator
- Kathy Zahul, Assistant State Traffic Engineer
- Derrick Cameron, Traffic Design Manager
- Phil Fergerson, Off-System Coordinator, District 6

Additionally, the Scan Team met with the following individuals:

- Lokesh Hebbani, FHWA Georgia Division
- Brandi Peck, Douglas County DOT
- Chris Pruitt, Cobb County DOT
- Talya Trudell, Atlanta Regional Commission
- Randy Clayton, Georgia Office of Highway Safety

### *Illinois*

The Scan Team met with several representatives from the Illinois Department of Transportation (IDOT) including:

- Priscilla Tobias, State Safety Engineer
- Darrel Lewis, Bureau Chief, Local Roads and Streets
- Rob Robinson, Data Management Unit Chief
- Kim Cummins, Local Roads Engineer
- Terry Fountain, Local Roads Engineer
- Traci Sisk, Special Programs Manager
- David Piper, Safety Design Engineer

Additionally, the Scan Team met with the following individuals:

- Jay Keeven, Commander, Statewide Patrol Support Command, Illinois State Police
- Rita Morocoima-Black, CUUATS Manger, Transportation Engineer, Champaign County Regional Planning Commission

- Alan Ho, Mobility & Safety Engineer Team Leader, FHWA Illinois Division
- Greg Piland, Safety Program Specialist, FHWA Illinois Division
- Meesa Otani, Transportation Engineer, FHWA Illinois Division
- Mike Staggs, Engineering Team Leader, FHWA Illinois Division
- Bob McCurdy, Lieutenant, Williamson County Sheriff's Office
- Brian Davis, Planning Engineer, Sangamon County Highway Department
- Tim Zahrn, County Engineer, Sangamon County Highway Department
- Jim Webb, County Engineer, Williamson County Highway Department

### *Michigan*

The Scan Team met with several representatives from the Michigan Department of Transportation (MDOT) including:

- Dale Lighthizer, Safety Programs Section Manager, Michigan DOT
- Tracie Leix, Safety Programs Unit Supervisor, Michigan DOT
- Dalrois McBurrows, Supervisor, Statewide Planning Section, Michigan DOT
- Jim D'Lamater, Local Agency Programs Safety Engineer, Michigan DOT

Additionally, the Scan Team met with the following individuals:

- Tom Bruff, Manager, Transportation Programs, South Eastern Michigan Council of Governments
- Wayne Schoonover, Ionia County Highway Engineer, Road Commission for Ionia County
- David Morena, FHWA Michigan Division
- Tim Colling, Assistant Director, Michigan Tech Transportation Institute, LTAP

### *Minnesota*

The Scan Team met with several representatives from the Minnesota Department of Transportation (Mn/DOT) including:

- Dave Engstrom, State Traffic Safety Engineer
- Julie Whitcher, Assistant Traffic Safety Engineer
- Mark Vizecky, Program Support Engineer
- Patti Loken, State Aid Programs Engineer
- Ken Schroepfer, Traffic Standards Specialist
- Kelvin Howieson, District 3 State-Aid Engineer
- Sue Groth, State Traffic Engineer

Additionally, the Scan Team met with Rick West, the County Engineer in Ottertail County, Minnesota.

***New Jersey***

The Scan Team met with several representatives from the New Jersey Department of Transportation (NJDOT) including:

- Patricia Ott, Director, Traffic Engineering and Safety
- William Beans, Section Chief, Bureau of Safety Programs
- Deven Patel, Principal Engineer, Division of Local Aid and Economic Development
- Debbie Kingsland, Project Engineer, Division of Planning and Project Development

Additionally, the Scan Team met with the following individuals:

- Pam Fischer, Director, New Jersey Division of Highway Traffic Safety
- Lois Goldman, Director, Regional Planning, New Jersey Transportation Planning Authority
- Jeff Vernick, Manager, Corridor Studies and Project Planning, North Jersey Transportation Planning Authority
- Tim Chelius, Executive Director, South Jersey Transportation Planning Organization
- Mark Jehnke, Principal Engineer, Ocean County Engineering Department
- John Ward, Associate Director, Planning Division, Delaware Valley Regional Planning Commission
- Janet Leli, Program Director, Center for Advanced Infrastructure and Transportation

- Sarah Weissman, Program Manager, Center for Advanced Infrastructure and Transportation and Transportation Safety Resource Center
- Caroline Trueman, Safety Engineer, FHWA New Jersey Division

***Washington***

The Scan Team met with several representatives from the Washington Department of Transportation (WSDOT) including:

- Kathleen Davis, Director, WSDOT Highways & Local Programs (H&LP)
- Matthew Enders, Technical Services Manager, WSDOT H&LP
- Aaron Butters, Engineering Services Manager, WSDOT H&LP

Additionally, the Scan Team met with the following individuals:

- Dan Davis, Collision Data & Analysis Branch Manager, WSDOT Transportation Data Office
- Don Peterson, Safety/Design Engineer, FHWA Washington Division
- Megan Hall, Local Programs/Research & T2 Engineer, FHWA Washington Division
- Jim Ellison, County Traffic Engineer, Pierce County Public Works
- Scott Davis, Traffic Engineer, Thurston County Public Works
- Angie Ward, Program Manager, Washington Traffic Safety Commission

## Scan Team Members

***Federal Highway Administration Representatives***

Rosemarie Anderson, Transportation Specialist, Local and Rural Roads Program, FHWA Office of Safety Programs, Task Order Manager

Jeffrey Shaw, FHWA Safety Engineer, FHWA Resource Center

Peter Hartman, Idaho Division Administrator, FHWA

***American Association of State Highway and Transportation Officials Representative***

Tracie Leix, Safety Programs Unit Supervisor, Michigan DOT

Representative of the AASHTO Standing Committee on Highway Traffic Safety (AASHTO-SCOHTS)

***American Public Works Association (APWA) Representative***

John Davies, Chief Engineer, Jacksonville Transportation Authority, Jacksonville, Florida

***National Association of County Engineers (NACE) Representatives***

Albert Letzkus, Chief Traffic Engineer, Pima County DOT, Tucson, Arizona

Jon Oshel, Director of Public Works, Tillamook County, Oregon

***National Local Technical Assistance Program Representatives***

Cliff Reuer, Technical Assistance Provider, South Dakota LTAP Center, South Dakota State University

John Habermann, Program Manager, Indiana LTAP, Purdue University School of Civil Engineering

***Vanasse Hangen Brustlin, Inc. (contractor to FHWA)***

Teresa Gardner, Project Manager

Manager, Consultant Services on the project.

Michelle Scism, Project Coordinator

***Additional Scan Contributors***

Karen Yunk, HSIP Program Manager, FHWA Office of Safety Programs

John Dewar, State and Local Programs Team Leader, FHWA Office of Safety Programs

Robert Hollis, Arizona Division Administrator, FHWA

Larry Dwyer, Vermont Division Assistant Administrator, FHWA

## End Notes

[1] FHWA Office of Highway Policy Information http://www.fhwa.dot.gov/policyinformation/pubs/pl10023/fig1_6.cfm

[2] MDOT's Time of Return (TOR) analysis measures the anticipated cost effectiveness of a project by using historic crash data to calculate the amount of money saved each year from crashes mitigated by the proposed project. Dividing the total project cost by the money saved each year in crash mitigation yields the number of years it will take to cover the costs of a project. The less time needed to recoup the investment, the more cost effective the project is considered. Projects that are cost effective will have a TOR that is less than 10

years long. (Genesee County 2035 LRTP, Safety Technical Report, http://www.co.ssgenesee.mi.us/gcmpc-plan/LRTPWeb/TechReports/Safety.pdf)

[3]As defined by Title 23 of the Code of Federal Regulations, "force account [is] the direct performance of highway construction work by a State transportation department, county...by use of labor, equipment, materials and supplies furnished by the agency and used under their direct control" [23 CFR 635 Subpart B, U.S. DOT FHWA. http://www.fhwa.dot.gov/legsregs/directives/cfr23toc.htm]

[4] U.S. DOT FHWA Legislation & Regulations, 23 CFR 771; http://www.fhwa.dot.gov/hep/23cfr771.htm#sec.771.117.

# INDEX

## A



## B



## C

## D



## E

## M



## N



## O



## P

## R



## S